PERGAMON INTERNATIONAL LIBRARY
of Science, Technology, Engineering and Social Studies
The 1000-volume original paperback library in aid of education, industrial training and the enjoyment of leisure
Publisher: Robert Maxwell, M.C.

INTRODUCTION TO BOND GRAPHS AND THEIR APPLICATIONS

OTHER TITLES OF INTEREST IN THE PERGAMON INTERNATIONAL LIBRARY

ANAND, D. K.
Introduction to Control Systems

BENSON, R. S.
Advanced Engineering Thermodynamics

BRADSHAW, P.
Experimental Fluid Mechanics, 2nd edition

BROOKES, A. M. P.
Basic Electric Circuits, 2nd edition

CRANE, P. W.
Electronics for Technicians

DIXON, S. L.
Fluid Mechanics, Thermodynamics of Turbomachinery, 2nd edition

DIXON, S. L.
Worked Examples in Turbomachinery (Fluid Mechanics and Thermodynamics)

GARNER, K. C.
Introduction to Control Systems Performance Measurements

GATLAND, H. B.
Electronic Engineering Applications of Two Port Networks

HANCOCK, N. N.
Matrix Analysis of Electrical Machinery, 2nd edition

HARRIS, D. J. and ROBSON, P. D.
The Physical Basis of Electronics

MORRILL, B.
An Introduction to Equilibrium Thermodynamics

PEERLESS, S. J.
Basic Fluid Mechanics

PRIDHAM, G. J.
Solid State Circuits

PRIDHAM, G. J.
Electronic Devices and Circuits, volumes 1, 2, and 3

"The terms of our inspection copy service apply to all the above books. Full details of all books listed and specimen copies of journals listed will gladly be sent upon request."

Cover: Hydraulic piston and slipper as an example of a hydrostatic bearing, schematically represented by a bond graph and a block diagram.

INTRODUCTION TO BOND GRAPHS AND THEIR APPLICATIONS

by

JEAN U. THOMA

Professor of Mechanical Engineering

PERGAMON PRESS

OXFORD · NEW YORK · TORONTO

SYDNEY · PARIS · BRAUNSCHWEIG

U. K.	Pergamon Press Ltd., Headington Hill Hall, Oxford OX3 0BW, England
U. S. A.	Pergamon Press Inc., Maxwell House, Fairview Park, Elmsford, New York 10523, U.S.A.
CANADA	Pergamon of Canada, Ltd., 207 Queen's Quay West, Toronto 1, Canada
AUSTRALIA	Pergamon Press (Aust.) Pty. Ltd., 19a Boundary Street, Rushcutters Bay, N.S.W. 2011, Australia
FRANCE	Pergamon Press SARL, 24 rue des Ecoles, 75240 Paris, Cedex 05, France
WEST GERMANY	Pergamon Press GMbH, 3300 Braunschweig, Postfach 2923, Burgplatz 1, West Germany

First edition 1975

Library of Congress Cataloging in Publication Data

Thoma, Jean Ulrich.
Introduction to bond graphs and their applications.

"Revised and rewritten by the author from the original German title Grundlagen und Anwendungen der Bonddigramme, Girardet Verlag, 1974."
Bibliography: p.
Includes index.
1. Bond graphs. I. Thoma, Jean Ulrich. Grundlagen und Anwendungen der Bonddigramme. II. Title.
TA338.B6T4613 1975 620'.001'5115 75-9763
ISBN 0-08-018882-6
ISBN 0-08-018881-8 pbk.

Printed in Great Britain by A. Wheaton & Co., Exeter

Contents

Preface

Bond graphs consist of signs, lines, and simple symbols for the representation of engineering systems. This book gives an introduction to the method with its simple conventions, and includes many applications from different engineering disciplines.

Bond graphs not only allow representation of systems on paper for self-study, simulation, and computation by hand, but they are at the same time a powerful tool for automatic computer programming with the ENPORT program described in Chapter 7.

The book addresses itself to all engineers interested in representing simple or complex engineering systems, especially systems with components from different engineering disciplines. It does not, with the possible exception of the last two chapters, require much mathematics, but a basic knowledge of control engineering would be helpful. It can be used as an introductory text for the different engineering disciplines, mechanics, fluid mechanics, and electronics with electromechanical power conversion or thermodynamics, if the students learn the few bond graph symbols at the same time. The uniform symbols and notation result in great simplicity and clarity especially if several engineering disciplines are covered.

Bond graphs were invented by Henry Paynter of MIT (Cambridge, Mass, USA) after intensive use of block diagrams for servocontrols and simulation problems. They were further developed by Dean Karnopp and Ronald Rosenberg, professors at the universities of California and Michigan respectively. The author met them when teaching a summer course on hydrostatic transmission and control systems at MIT, which was based on his earlier book *Hydrostatic Power Transmission,* Trade and Technical Press (1964) with many block diagrams, of which bond graphs are a natural extension.

Continuous contact, visits, and seminars with Paynter, Karnopp, and Rosenberg and teaching in the United States and Continental Europe allowed elaboration of the many applications included in this book. The author feels that bond graphs are a simulation and representation tool that maintain, unlike some other tools, the essential physical principles and laws which are the fundamental reason for their power. In fact, bond graphs are also most suitable for expounding many basic effects of physics and should therefore become increasingly useful in all disciplines of engineering science.

Zug, Switzerland Jean U. Thoma
Bellevueweg 23
Spring 1975

Introduction

Bond graphs display both energy and signal exchanges between components or elements in installations and systems by simple lines and symbols. They bridge the gap between control engineering and the parts of engineering science where power and efficiency have greater importance including energy conversion.

The book shows how many fields of engineering science can be described by the few standard symbols of the bond graphs thoroughly and yet in an intuitively appealing manner. This indicates the common structure of all our engineering concepts only too frequently masked in textbooks by different notations.

Apart from the description of technical systems, bond graphs allow the systematic setting up of block diagrams of control engineering. Furthermore they allow programming a computer directly and automatically with the ENPORT programs.

Chapter 1 çompares the different kinds of graphs, diagrams, and models as a preparation for the fundamentals of bond graphs in Chapter 2 with tables for the standard elements.

Whilst the fundamentals of bond graphs are easily grasped, their elegance and power follow from the broad range of applications. Thus Chapter 3 introduces their systematic application with mechanical examples, Chapter 4 brings fluid power engineering (sometimes called oil hydraulics), and Chapter 5 electrotechnique and electronics. Chapter 6 deals with thermodynamics, where entropy is the fundamental variable as a displacement. In this discipline especially, (difficult to many), bond graphs contribute much to a clearer exposition. Chapter 7 treats simulation and electronic data processing directly from a bond graph with the ENPORT program, leading to an important reduction in the difficulty of programming.

Chapter 7 brings additional material from the different disciplines and requires, together with Chapter 8, perhaps a little more mathematical knowledge from the reader. Apart from the explanation of magnetic circuits, actuator magnets, and transformers with stray flux, it contains the thermodynamics of flow processes (irreversible thermodynamics) which can be described very clearly by the standard elements of bond graphs.

At the end of this book there is a small glossary with some important technical terms in English, German, and French. The Appendix contains a short table of the most important symbols as an aid to memory in addition to the extensive tables of Chapter 2.

CHAPTER 1

Diagrams, Models, and Simulation

1.1. Graphs and Diagrams as Representations of Engineering Systems

Bond graphs are a representation of components, machines, and systems on paper using conventional signs and symbols as described in detail in Chapter 2. Here the different possible representations of systems on paper will be described generally in order to obtain a good basis for the introduction of bond graphs. Representations of many kinds are used frequently in engineering science, starting from simple components or machine assemblies to complex mechanical, electrical, and chemical installations called comprehensively Engineering Systems. Such representations are needed not only for the conceptual and detail design of technical systems, but also for their sale and correct operation and maintenance.

It is very useful to have a clear idea about the different representations with their own rules and conventions. As shown in Fig. 1.1, they can be in word form for which the user must frequently learn new technical terms. Alternatively, technical systems can be represented by figures and drawings, including especially the standard engineering drawings with their special symbols and conventions.

Many technical systems are usefully considered as consisting of single elements interacting through connections. They can be represented (Fig. 1.1, right) by circuits or a little more abstractly by diagrams. Circuits include the circuit diagrams of electronics and of fluid power engineering with their internationally agreed symbols. Such circuits are also used for building an engineering system from given components.

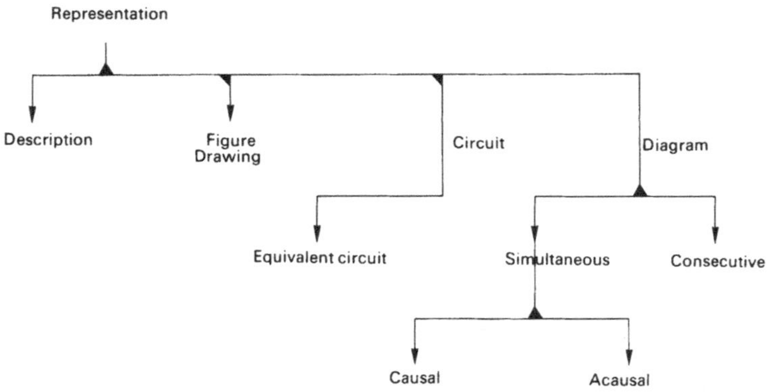

Fig. 1.1. Family tree of representations of engineering systems.

A variant of circuits are the equivalent circuits described below. It should be noted that the conventional circuit symbols consist of more or less idealized figures or pictures of the elements or components.

Diagrams are either simultaneous, where all parts work at the same time, or consecutive, where a representative point travels through the diagrams and indicates the instantaneous state of the system. Such consecutive diagrams include the flow charts used for computer programming and also the known PERT and CPM charts.

Simultaneous diagrams are either causal, that is with a certain direction of action between the elements, or acausal without such a distinction between cause and effect. The important concept of causality will be described in more detail below (§ 1.3) with the block diagrams of control engineering.

As the subject of this book, bond graphs are simultaneous diagrams in the sense of Fig. 1.1 set up acausally and transformed subsequently into a causal diagram by a systematic choice of the causalities. This step can also be done by the computer (Chapter 7).

In practice the different representations are used frequently together as a descriptive test illustrated by drawings, circuits, or diagrams. Diagrams should be simple and time-saving to draw, their symbols being highly simplified pictures or figures of the real components. The more, however, a symbol is simplified, the more difficult is it to memorize and to use for qualitative considerations. Consequently there is an optimum to simplification beyond which the representation becomes too abstract.

1.2 Models for Technical Systems

The different representations of an engineering system are intended to show how it operates and to indicate the value and time dependence of its different physical variables. This provides the necessary information for a detailed design of the components. In short, the engineer desires a model of the system.

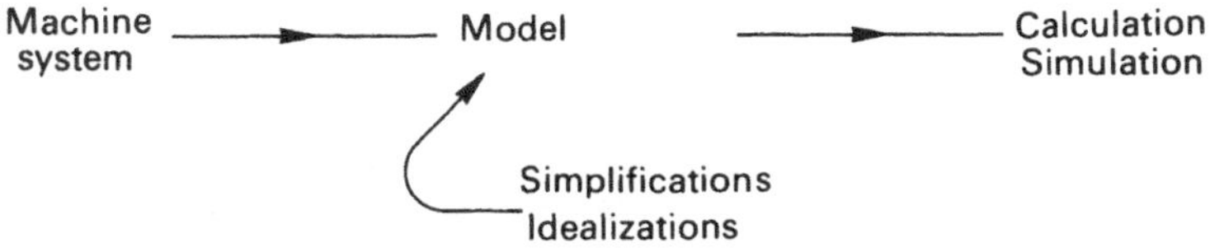

Fig. 1.2. Relations between system, model, and simulation.

As shown on Fig. 1.2, a model of a system contains the essentials of its operation under the laws of physics, where unimportant influences are neglected and real processes are idealized. With the model the values of the variables can be calculated or the system simulated, which means calculation of the behaviour of the variables on a computer. Another important use for a good model is the estimation of the system behaviour with the maximum values of the variables from the laws of physics.

Several models can be set up for each system according to the required accuracy and to what questions are important. More accurate models are unfortunately more complex, leading to more expensive programming and operating of computers.

Setting up a good model for a system or a process is an important task of the engineer since it can considerably facilitate the calculations. Circuits, equivalent circuits, and diagrams of any kind are models in this sense, and also engineering drawings can be considered as models, especially if they are drawn somewhat schematically.

In a larger sense even a system of equations or a theory about the operation of a system is an abstract model.

The well-known mathematical models are in principle certain ideas about the operation of a system, usually a component or a subgroup (e.g. a hydrostatic machine, Chapter 4), and also as the starting point for a theory. Quantitative indications and numerical values can be added and the result can be given as table, graphical representation, or mathematical formula. Mathematical models in the strict sense are formulae set up from theoretical considerations where the value of the constants is determined by experiments. Such models are very useful for simulating components in a larger system.

1.3. Block Diagrams of Control Engineering and their Inversion

The block diagrams have grown from the need of control engineering to divide a system into components with simple behaviour represented by blocks. It should be noted that the connections between the blocks contain only one variable running in the direction of the arrow. In contrast a real connection always contains two variables, as, for example, current and voltage in an electric wire.

Each block of a block diagram is without return effect on the preceding block as is usually true for real control systems. As an example, the actuator of a control valve of a steam turbine is sufficiently powerful for the flow forces of the valve to have no influence on the position of the actuator which is given only by the input signal.†

The operation of the components in block diagrams is idealized in the sense that the blocks contain only a functional dependence between input and output variables or input and output signals. This dependence can be linear or non-linear, instantaneous or time-dependent, and each block has only one output connection but can have several input connections. With time-dependent blocks the out-

†The signal flow graphs of control engineering are essentially the same representation with more abstract symbols (ref. 4).

put signal is often the derivative or the time integral of the input signal. It is then represented on the block by s or 1/s indicating the derivation or integration operator. In linear blocks the output signal is proportional to the input signal (by definition). Their ratio is independent of the value of the signal but a function of the signal frequency in time-dependent blocks. It is the gain of the block and is usually denoted by a letter inside the block.

The block diagrams of control engineering are built from the following elements:

(1) Blocks with a functional dependence between input and output signals.
(2) Connections where the signal travels in the direction of the arrow.
(3) Branch points where the signal enters several connections without changing its value.
(4) Mixing points (addition and subtraction points), where the output signal is the sum of the difference of the input signals.

Figure 1.3 contains a mixing point (left) and a branch point (right) of the block with the gain F.

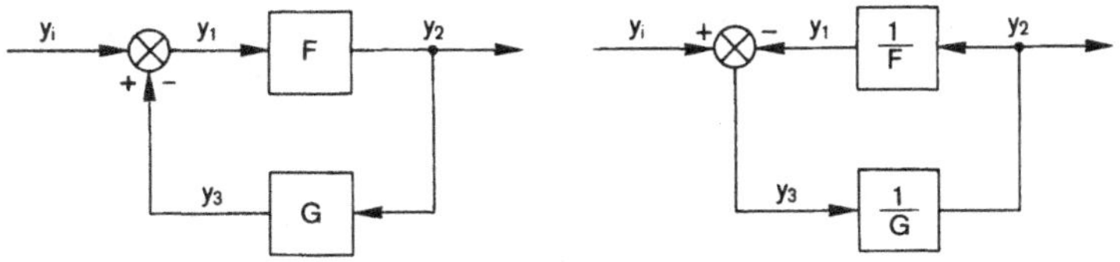

Fig. 1.3. Simple block diagram (left) and its inversion (right).

Since block diagrams prove to be useful especially for time-dependent signals, they have been applied also for real components with different internal return actions shown by suitable feedback connections. In this case the single blocks represent not a component but a function. †Similarly, any return action between components can be represented by separate feedback connections.

†This is the reason why the author has used the term functional diagram in ref. 5.

Figure 1.3 contains (left) a simple block diagram with the input variable y_i, the error variable y_1, the output variable y_2, and feedback variable y_3. They are connected by the following equations:

$$y_1 = y_i - y_3 \qquad y_2 = Fy_1 \qquad y_3 = Gy_2. \tag{1-1}$$

The product of the gains of both blocks is called the (open) loop gain, since it represents the gain experienced by a signal travelling around the loop. The ratio between input and output variables, i.e. the transfer function of the system (with closed loop) is obtained from (1-1) by eliminating the intermediate variables:

$$\frac{y_2}{y_i} = \frac{F}{1 + FG}. \tag{1-2}$$

The transfer function is seen to be equal to the gain of the block in the forward connection divided by one plus the loop gain. This is often called the formula or Mason's rule.

It is possible to invert the diagram (Fig. 1.3, right) whence the signal runs through the blocks in reverse and the block gains become reciprocal. This corresponds to the following equation:

$$y_3 = y_i - y_1 \qquad y_2 = \frac{1}{G} y_3 \qquad y_1 = \frac{1}{F}\ y_2. \tag{1-3}$$

The loop gain is now $1/FG$, but the transfer function remains the same as seen from Mason's rule:

$$\frac{y_2}{y_i} = \frac{\frac{1}{G}}{1 + \frac{1}{FG}} = \frac{F}{1 + FG}. \tag{1-4}$$

The movement from the left to the right part of Fig. 1.3 represents the inversion of a block diagram. The block gains and the loop gain become reciprocal. In particular, if in one form the loop gain is large compared to one, it is small in the inverted block diagram. The loop gain is dimensionless in the sense of the theory of the dimensions of physical variables. In complex block diagrams with several closed loops, each loop can be inverted.

The inversion may appear strange to many control engineers since the real control components are usually not reversible. Whilst the admissibility of inversion is treated in ref. 4, it should be noted here that as soon as the blocks represent functions and not components, there is no basic obstacle to inversion.

Inversion has the following effects:

(1) Linear blocks receive the reciprocal gain.
(2) In non-linear blocks the inverse function must be entered.
(3) In time-dependent blocks integration is replaced by derivation and derivation by integration.†

Some practical difficulties may arise with many valued functions treated with further information about diagrams, block diagrams, and inversions in ref. 4.

The signals travelling in the sense of the arrows define causes and effects. For instance, in the left part of Fig. 1.3 and in eqn. (1-1), y_1 is the cause of y_2, and y_3 the effect of y_2. Consequently the block diagram is a causal representation in the sense of Fig. 1.1. Contrariwise, in the right part of Fig. 1.3 and in eqn. (1-3), y_3 is the cause of y_2 and y_1 the effect of y_2. Thus the right part of Fig. 1.3 is also a causal diagram but with different causality.

The determination of causes and effects constitutes the causality of a block diagram. The inversion described above is really the transition from one causality to another which considerably changes its appearance of a block diagram, especially in complex cases.

Before a block diagram can be drawn, the causalities must be selected. They are not given by nature but chosen by the engineer for his analysis. This artificial character is a disadvantage of all causal diagrams and the computer programs based on it.

It is interesting to note that many of our laws of physics are taught in a certain causality. Mathematical equations contain in principle a causality from right to left, since the independent variable is written conventionally at the right-hand side and the dependent variable at the left-hand side.

A definite causality is also implicit in electromagnetism, where a change of the magnetic field induces a voltage which in turn produces a current depending on the electrical resistance. This way of looking is well adapted to

† A block with a time delay would be replaced by an advance operator, but this is physically unrealizable.

normal electrical effects, but not to superconducting circuits that continuously become more important. Here the electrical resistance vanishes and the gain between induced voltage and electric current is infinitely large if the accelerating force of the electrons is neglected. Consequently, with superconducting wires the sum of the current in a loop and of the intersecting magnetic flux (better as magnetic displacement according § 8.4) is constant.

In spite of their usefulness in control engineering, block diagrams have the following drawbacks:

(1) A return action between components must always be shown by a separate feedback connection, whilst in reality the components have only one connection such as electrical wires, pipes, or shafts.
(2) Block diagrams for a given technical system can have a radically different appearance because of the different possible causalities.

Bond graphs are introduced to avoid these drawbacks of block diagrams.

1.4. Circuits and Equivalent Circuits

Since circuits and equivalent circuits are frequently a useful alternative to bond graphs, they will be described here as another method.

Circuits are composed of element or circuit symbols which are more or less simplified or stylized pictures of the real components. The connections, e.g. electrical wires or pipes, between them are usually represented by simple lines. Internationally agreed symbols exist in many disciplines for the components and the connections, as in electronics and in fluid power. On the other hand, universal symbols for simple mechanical components, e.g. shafts, brakes, and torsional springs, are still lacking.

Similarly, as with block diagrams, there is the tendency to assign to each element of a circuit only one function, i.e. to take it as an ideal element. The characteristic of real components can be represented by added fictitious elements. A familiar example in electronics, and in fluid power is a real voltage (or pressure) generator shown as an ideal voltage source with a resistor in series, the so-called internal resistance. It well represents the characteristics between voltage and current of a real generator (Fig. 4.16).

The added fictitious elements transform the real circuit into an equivalent circuit. This method is normally used in electronics, where the ideal voltage source (with a voltage independent of current) and the ideal current generator (current independent of opposing voltage) are important elements.

As an example of fluid power, Fig. 1.4 shows a hydrostatic pump by its AHEM symbol in the United Kingdom (equal to the NFPA symbol in the United States and to the German symbol according to DIN 24 300), where the unavoidable leakage returns to the lower tank as shown by the broken line. The right-hand side of Fig. 1.4 contains an equivalent circuit where the pump is considered ideal or loss free, and the leakage is represented by the parallel resistor and the torque losses by the symbol for a disc brake.

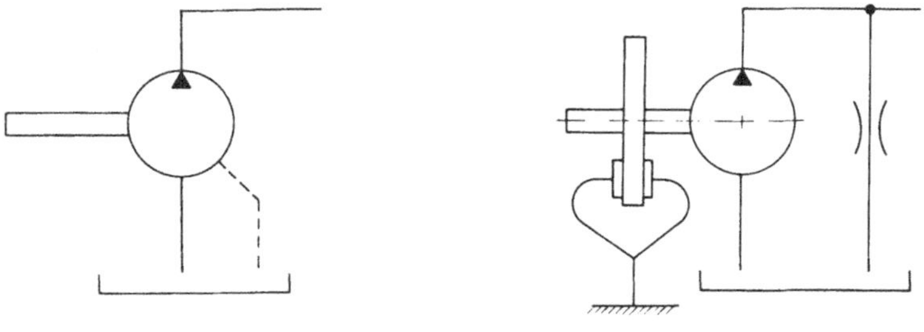

Fig. 1.4. Circuit symbol of a hydrostatic pump (left) with added brake and hydraulic resistor for the losses (right)

Equivalent circuits are widely used, especially in electronics in spite of their following drawbacks:

(1) Due to the fictitious elements the equivalent circuit loses its similarity with the real layout. It can no longer be given to a student or fitter to build an installation from available components since the fictitious elements have no hardware equivalent.

(2) With each symbol it must be indicated separately, whether it is a real component or an idealized element. Especially outside electronics this distinction is not universally recognized.

(3) New symbols must be learned for each discipline (electronics, fluid power, and mechanics), and in mechanics especially many different symbols are in use as mentioned above.

Bond graphs avoid these drawbacks of equivalent circuits and can represent interdisciplinary systems with one set of symbols for all disciplines.

Instead of bond graphs, many authors use the so-called network graphs or linear graphs, mentioned briefly in § 2.1 below. They are simultaneous diagrams in the sense of Fig. 1.1, written at first acausally, whence the causalities are selected by dividing the graph into trees and branches. The network graphs, first introduced in electrotechnique, have some common properties with bond graphs but are intuitively much less appealing and less useful for interdisciplinary systems. For an excellent introductory comparison, see ref. 22.

CHAPTER 2

Fundamentals of Bond Graphs

2.1. Bond Graphs with Words for Components

Having discussed the different representations of engineering systems in Chapter 1, this chapter deals systematically with the fundamentals of bond graphs. Bond graphs are based on the splitting of engineering systems in separate components that exchange energy or power through identifiable connections or ports. They are called bonds in analogy to the energy exchange between atoms in chemical bonds. With this in mind it is possible to set up bond graphs first summarily, i.e. in rough lines and to refine them progressively until they are a sufficiently accurate model of the system as discussed in § 1.2.

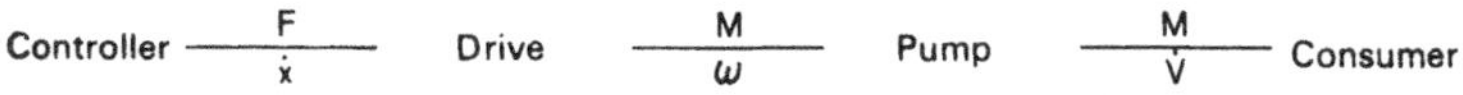

Fig. 2.1. Word bond graph for a pump drive.

Bond-graph elements can be represented initially by words and the bonds for power exchange between them as simple lines. Figure 2.1 contains, a variable drive, e.g. a Diesel engine with a pump that supplies fluid to a consumer. The energy flux or power in a bond is always the product of two variables – a potential variable called effort and a flow or current variable, simply referred to as flow.

Table 2.1 shows the usual effort and flow variables with the conventional letters for mathematical formulae, in particular electric voltage and current, pressure and volume flow (for fluid power engineering with incompressible fluids), force and velocity, or torque and rotation angular frequency (also called rotative speed) in

Table 2.1. *Effort and flow variables from different variables*

	Electro-technics	Hydraulics (incompressible)	Mechanics (displacement)	Mechanics (rotation)	Thermo-dynamics	General
Effort variable	Electric voltage u	Pressure p	Force F	Torque M	Absolute temperature T	e
Flow variable	Electric current i	Volume flow $\dot{V}$	Velocity v or $\dot{x}$	Rotation frequency ω or n	Entropy flow $\dot{S}$	f
Displacement	Electric charge q	Volume V	Displacement x	Rotation angle φ	Entropy S	q
Momentum	Voltage pulse	Not named	Momentum p	Angular momentum L	Not named	p

mechanics. Finally, there is the absolute temperature and the entropy flow in thermodynamics (Chapter 6). In general, the effort is denoted by e and the flow by f.

Also shown in Table 2.1 are the displacements or time integrals of the flows, including electrical charge, volume, displacement or angle of rotation, and entropy in their respective disciplines. Finally, the generalized momentum as time integral of the effort variable has been included, although it is only normally used in electronics as voltage pulse and in mechanics as momentum or angular momentum.

As described more fully in § 8.6, it would be possible to identify force and torque as flow and velocity, and rotation frequency as effort. This can be referred to as the dual analogy as opposed to the direct analogy of Table 2.1. The dual analogy has the advantage that it maintains the structure of the network graphs mentioned briefly in § 1.4. There is nothing wrong with it mathematically, but for the practical engineer a flow is always associated with movement, and this is the reason why the author and most specialists of bond graphs prefer the direct analogy. Section 8.6 contains a comparison of both analogies.

The symbols for the effort are written above the bonds and below for the flows (respectively right and left with vertical bonds), as shown in Fig. 2.1. Here f is the force and x (instead of v) the velocity of the control rod for changing the fuel injection of a Diesel engine.

It should be noted that the word-bond graph of Fig. 2.1 has still a certain similarity with a circuit of the installation. The main difference is that the components are designated by words instead of figures or symbols. The similarity is lost when the words are replaced by the standard elements of bond graphs described below.

The fluid delivered by the pump to the load must return through a separate conduit or through an open common tank. In Fig. 2.1 it is supposed implicitly that the fluid returns without appreciable pressure and therefore no energy transport is associated with it. Similarly, the torque must be returned by the reactions between the drive and the pump, but this also transports no energy since the case is immobile. The return transport becomes important only with moving bases or reference frames, e.g. with planetary gears (§ 3.3) and with series circuits of electrical and hydraulic components.

Bond graphs are useful for interdisciplinary systems since the energy transport is generally the product of an effort and a flow variable. Also important, apart from the product, is the ratio of effort and flow referred to as impedance with its reciprocal, the admittance. With the energy flux $\dot{E}$, the impedance Z, and the admittance Y, we have

$$\dot{E} = ef \qquad Z = \frac{e}{f} \qquad Y = \frac{f}{e}.$$

In words:

- The electrical impedance is the ratio of voltage and current.
- The hydraulic impedance is the ratio of pressure and volume flow.
- The (rotative) mechanical impedance is the ratio of torque and rotation (angular) frequency and the (linear) mechanical impedance the ratio of force and velocity.
- The thermal impedance is the ratio of absolute temperature and entropy flow.

The thermal impedance is used very rarely but might become useful.

Each bond is thus characterized not only by the energy flux but also by the impedance as an alternative to the indication of effort and flow. The concept of impedance refers consequently not only to the resistance elements of § 2.2, but generally denotes the ratio of effort and flow in bonds.

2.2. Standard Elements of Bond Graphs

The next step from a bond graph with words for the components is to introduce the standard elements and appropriate connections and junctions. In practice only two kinds of junctions and very few standard elements are needed to model most engineering systems accurately.

The standard elements of bond graphs are classified according to the number of bonds as one-ports, two-ports, and multiports. This classification originates from electric circuit theory where each port or bond represents two wire terminals or connections. A one-port has two connections, a two-port four connections replacing progressively the older designations dipole and quatripole. Refering to Fig. 2.1, the controller and the load are one-ports (with only one bond) and the drive and the pump two-ports with two bonds each. Apart from this, there are elements with three or more bonds called three-ports and multiports. We shall now describe the standard elements of bond graphs in this order.

2.2.1. *One-ports*

One-ports are elements exchanging energy with the system through one bond only. They include resistance elements, called R-elements; inertia elements, called I-elements; and capacity elements called C-elements. In addition there are the effort and flow sources known from electronics. Table 2.2 shows a simple bond (Top), then the activated bond described below and one-ports with examples from electronics, fluid power engineering, and mechanical engineering. In addition to the bond-graph symbols the table contains the usual circuit symbols in the different disciplines and their equations.

The R-elements generalize the electrical resistances and have a functional relation between effort and flow. This relation is linear in a linear R-element, but in general it is non-linear, e.g. the quadratic relation between pressure drop and volume flow in turbulent fluid flow. It should be particularly noted that the R-element withdraws power from the system and transforms it mostly into useless heat (Chapter 6). Consequently, the unavoidable losses of real components and machines are produced by undesired R-elements.

The I-element corresponds to the electrical inductivity and to the action of masses in mechanics. It gives a usually linear relation between the time derivative of the flow and the effort, or, in other words, between flow and the integral of effort – the generalized momentum. A well-known example from mechanics is the flywheel with its moment of inertia. An inertia element does not dissipate energy but stores it. When it is brought back to its original state, e.g. when the flywheel is slowed down it returns the entire energy to the system.

Table 2.2. *Standard elements of bond graphs: one ports*

Symbol	Designation	Examples		
		Electrotechnics	Fluid power	Mechanics
$\frac{e}{f}$	Simple bond equation $\dot{E} = ef$	Electric wire $\dot{E} = ui$	Pressure pipe $\dot{E} = p\dot{V}$	Shaft or rod $\dot{E} = M\omega = Fv$
⟶	Activated bond			
One-ports				
$\frac{e}{f}$ R	Resistance element R-element $e = Rf$ linear $e = F(f)$ non-linear	 $u = Ri$	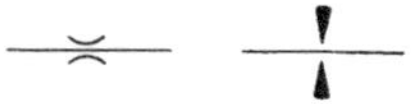 $p = Z\dot{V}$	$M = Z\omega = Zv$
$\frac{e}{f}$ I	Inertia element I-element $e = I\frac{df}{dt}$ $p = \int e\,dt = If \cong F(f)$	 Inductor $u = L\frac{di}{dt}$	No symbol Mass action $p = I\frac{d\dot{V}}{dt}$	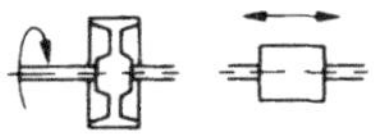 Flywheel, mass $M = J\frac{d\omega}{dt}$
$\frac{e}{f}$ C	Capaçity element C-element $f = C\frac{de}{dt}$ $q = \int f\,dt = Ce$	 Condensor $i = C\frac{du}{dt}$ $q = Cu$	 Hydraul. accumulator $V = Cp$	 Linear or torsional spring $M = k\varphi$ $\omega = \frac{1}{k}\frac{dM}{dt}$

S_e ——	Effort source Effort independent of flow		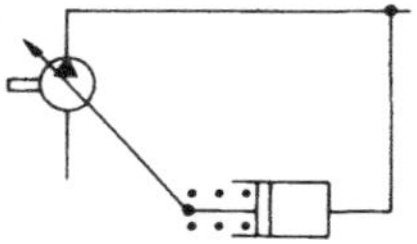	No symbol
⟶ S_e ——	Controlled effort source	Voltage source	Hydrostatic pump with pressure regulator	Diesel engine
⟶ S_f ——	Flow source Flow independent of effort	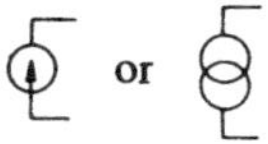		No symbol
		Current source or generator	Hydrostatic pump	Transmission shaft driving many machines

The C-element has a relation between the time derivative of effort and the flow or between the effort and the time integral of the flow, which is the displacement. Known examples are the mechanical spring and the electrical capacitor. These elements are usually linear, whilst, for example, the hydropneumatic accumulator of a fluid power is a non-linear C-element. With this accumulator the pressure is a curved function of the absorbed volume according to the gas laws. Also C-elements do not dissipate but store the energy and return it on discharge.

Other one-ports are the effort and flow sources known from electronics, which usually deliver energy into the system. In an effort source, the effort is independent of the flow drawn, and in a flow source the flow is likewise independent of the opposed effort. In a real component this is, of course, only the case as long as this opposed effort is not too large.

An example of a flow source from fluid power is a hydrostatic pump driven with constant rotation frequency, where the volume flow is practically independent of the opposed pressure. If such a pump is equipped with an ideal pressure regulator it becomes an effort source with the effort (the pressure) independent of flow. The losses of real pumps are represented by R-elements (Chapter 4) and the delays of real pressure regulators by I- and C-elements.

Many effort flow sources are controllable by a signal as shown by a line with a double arrow. Similarly a double arrow touching an R-, I-, or C-element shows that its value is adjustable, e.g. the capacity of an electric capacitor. Such a line with a double arrow is an activated bond as described in § 2.3.1.

As with all models it is necessary to indicate for R-elements and effort sources for which range of variables the functional relation is valid. With normal R-elements the flow can reverse its direction or become negative, whence the effort also changes its direction. Consequently the direction of energy transport as product of flow and effort remains the same. It runs always into the R-element and is dissipated there, illustrating the fact that the dissipation is irreversible, as frequently described in thermodynamics (Chapter 6). With quadratic resistance elements (turbulent fluid restrictors), the effort or pressure drop changes its sign with the flow. Strictly speaking, the pressure drop is not proportional to the square of the flow but proportional to the flow multiplied by the absolute value of the flow.

In most effort sources (similar relations hold for flow sources) the effort is unchanged if the flow reverses its direction and enters into the source. Then the energy transport changes its direction and enters the source; this is later referred to as a reverse driven source (Chapters 4 and 6). Many but not all sources can be driven in reverse, and the validity of this model must be ascertained in each case.

2.2.2. *Two-ports*

Two-ports have two bonds for energy exchange with the system. They are energy conserving in the sense that the product of input flow and effort is at all times equal to the product of output flow and effort. As shown in Table 2.3 as symbols with different examples, two-ports are:

(1) The transformer (denoted by TF), where both efforts and both flows are proportional due to the power conservation. The ratio of the efforts is called the transformer ratio (or modulus) and is equal to the reciprocal ratio of the electric currents or flows. Other examples for transformers are pumps and cylinders in fluid power engineering, and levers and gear reducers in mechanical engineering.

(2) The gyrator (over-crossed transformer and denoted GY), where the input effort is proportional to the output flow and the output effort proportional to the input flow. The constants of proportionality are again equal due to power conservation and are called gyrator ratio (or modulus). A well-known example is the ideal electrical direct-current motor, where the torque multiplied by the gyrator ratio is the armature current and the armature voltage so multiplied is the rotation frequency. Consequently the product of torque and rotation frequency is equal to the product of voltage and current and is power conserved (see Table 2.3).

Both transformer and gyrator may be adjustable, and their

Table 2.3. ***Standard elements of bond graphs: two ports***

Symbol	Designation	Examples		
		Electrotechnics	Fluid power	Mechanics
— TF —	Transformer $e_2 = \lambda e_1$ $f_1 = \lambda f_2$ Note: Frequently, especially in mechanics, the letter i is used (instead of λ) for the transformer ratio	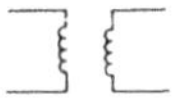 $u_2 = \lambda u_1 ; i_1 = \lambda i_2$ Electric transformer	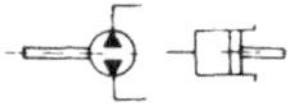 $M = V_0 p; \dot{V} = V_0 \omega$ Hydrostatic pump and cylinder	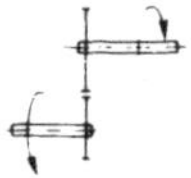 Gear reducer $M_2 = iM_1 ; \omega_1 = i\omega_2$
— GY —	Gyrator (overcrossed transformer) $e_2 = mf_1$ $e_1 = mf_2$	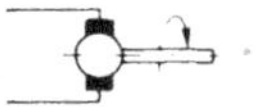 $M = mi$ $u = m\omega$ Electric DC machine with constant excitation	—	—
λ — MTF —	Modulated transformer	Variable transformer	Variable pump	Transmission with adjustable ratio
m — MGY —	Modulated gyrator	Electric DC machine with adjustable excitation	—	—

conversion ratios can be modulated by a signal applied through an activated bond. This is represented in Table 2.3 by the symbols MTF and MGY with the modulating parameters λ and m. Examples are variable ratio electric transformers, variable displacement pumps, and electric motors with variable field current.

Transformers, although they are not normally distinguished in bond graphs, are really of the following two different kinds:

(1) The impedance transformers, where the input and output variables are of the same energy domain or class. They change the impedances, i.e. the ratio of effort to flow in the bonds, e.g. electrical transformers or gear reducers.

(2) The class transformer, where input and output variables belong to different energy domains or classes (disciplines), e.g. hydraulic cylinders and pumps connecting hydraulic with mechanical variables. Since the impedances here belong to different classes (have different physical dimensions), it is not possible to compare them.

Class transformers are important since they allow an efficient representation of interdisciplinary engineering systems by bond graphs. It should be mentioned again that both class and impedance transformer maintain the power at all times.

Some American publications use the symbol TD (transducer) for the class transformer. We have not adopted this usage since some signal transducers are really gyrators (§ 5.3).

2.2.3. *Junctions*

The elements of bond graphs are connected by junctions of which there exist really only two kinds as shown in Table 2.4:

(1) The p-junction, where the efforts are equal and the flows additive or, more precisely, add up to zero (§ 2.3). This junction corresponds to the parallel circuit in electronics and in fluid power and is represented in mechanics by levers and planetary gear reducers (§ 3.3).

(2) The s-junction, where the efforts add up to zero and the

Table 2.4. *Standard elements of bond graphs: junctions*

Symbol	Designation	Examples		
		Electrotechnics	Fluid power	Mechanics
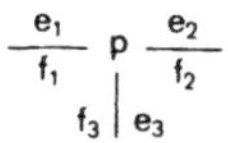	Parallel junction p - junction $e_1 = e_2 = e_3$ $f_1 + f_2 + f_3 = 0$	 Parallel circuit	 Parallel circuit	No symbol Connection through planetary gear or compensating levers
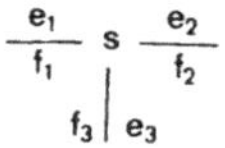	Series junction s - junction $e_1 + e_2 + e_3 = 0$ $f_1 = f_2 = f_3$	 Series circuit	 Series circuit	No symbol Connection through rigid couplings or levers

flows are equal. It corresponds to the series circuit in electronics and fluid power and to fixed couplings and connections in mechanics.

Junctions are power conserving at each instant and the power transports of all bonds add up to zero at all times. The example junctions of Table 2.4 have three bonds each (also called three-ports). Obviously there exist junctions with four or more bonds corresponding to the parallel or series circuits of several components. Such junctions are an example of multiports.

Some authors finally use a general junction where only the power flux is conserved, but the efforts and flows are arbitrarily connected by using the symbol jct. We shall use such a general junction for the energy conversion by C-converters in Chapters 5 and 6. There occasionally we shall also need a time domain junction indicating that different bonds are connected consecutively.

When setting up bond graphs systematically (§ 3.3) junctions with only two bonds will appear. As shown in § 2.4, such junctions give an identity and can be eliminated immediately since both efforts and both flows are equal. This is strictly true only with proper choice of the signs such that the positive energy flux enters the junction through one bond and leaves it over the other bond.

Finally, it should be noted, that many American publications use the designation o- and l-junction instead of p- and s-junction. This notation is more difficult to memorize and less favourable for computer programming.

The Appendix of this book contains a short table of the most important symbols of bond graphs including the fields of § 2.4.

2.3. Augmentation of Bond Graphs

After replacing the words by the standard elements of § 2.2 a bond graph is further built by selecting the direction of the positive energy flux and the causalities, i.e. the determination of causes and effects. This is called augmentation or augmenting of bond graphs.

2.3.1. *Causality and activated bonds*

Although already discussed in § 1.3, the important concept of causality is explained once more by the example of the simple installation with controller, drive, and load of Fig. 2.1. It can be described by saying that the drive produces a torque acting on the load which reacts by the rotation frequency on the drive. This is shown in Fig. 2.2 as a bond graph (top) and as a block diagram (bottom), and represents a possible choice of causality. In the bond graph it is denoted by the cross-stroke on the right indicating that the effort acts to the right, the side of the cross-stroke and the flow in the reverse direction, to the left. In the block diagram the direction of action is indicated by an arrow on each connection.

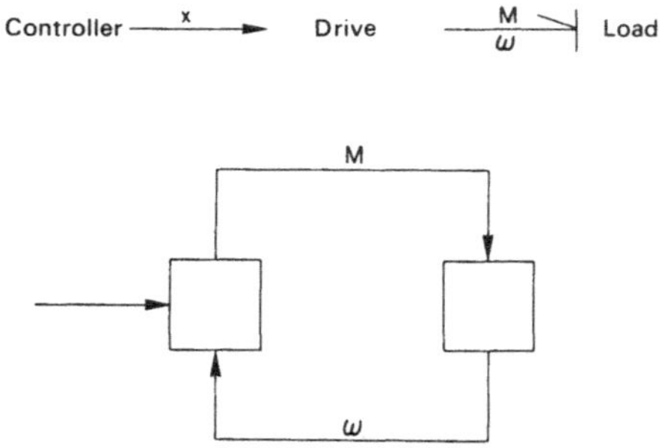

Fig. 2.2. Bond graph and block diagram of a drive with causality.

The operation can be interpreted alternatively according to Fig. 2.3, where the drive acts with the (angular) rotation frequency ω on the load that opposes a torque M. This is a different causality indicated on the bond graph of Fig. 2.3 (top) by the cross-stroke on the left. Consequently the effort acts to the left and the flow to the right. The bottom part of Fig. 2.3 shows the block diagram in this causality.

A change of the causality is indicated on the bond graph by a different position of the cross-stroke, whilst in the block diagram it is given by the inversion theorem (§ 1.3).

Sometimes only one variable of the bond is effective or active (as shown in Fig. 2.1, velocity or the position of the control rod), whilst the return action of the force of the controller can be neglected.

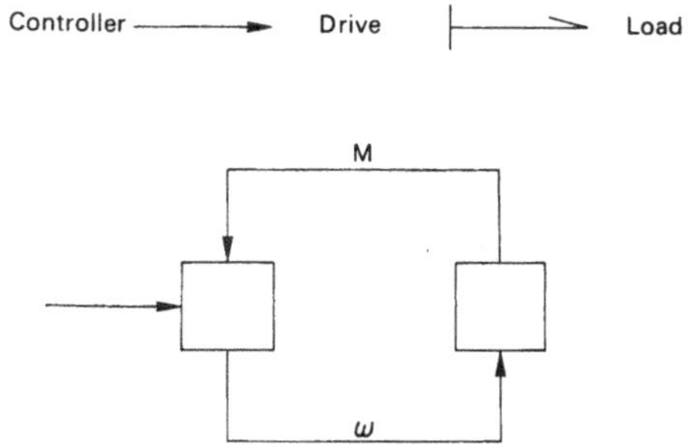

Fig. 2.3. Bond graph and block diagram of a drive with different causality.

Consequently only one variable is important as shown in Figs. 2.2 and 2.3 by the arrow. Such a connection is called an activated bond and gives the transition to the block diagram which in this sense contains only activated bonds.

Bond graphs with several elements correspond to block diagrams with several closed loops; the causality of each can be chosen. This will be described with the applications, and § 2.3.3 gives some rules about the choice of the causality which is otherwise free.

Activated bonds originate from normal bonds usually over junctions. Here the convention is valid that an activated bond starting from a p-junction contains the common effort and an activated bond starting from a s-junction contains the common flow, as we shall use frequently later on.

In addition to the activated bonds which really are an element of block diagrams (§ 1.3) any other element including these same block diagrams can be used in bond graphs. This specifically includes the integrating block.

2.3.2. *Choice of positive direction of energy flux*

Independently of the causality, the direction of the positive energy flux must be chosen for each bond. This is equivalent to the determination or convention of signs, and is indicated by a half-arrow, see for example Figs. 2.2 and 2.3. The choice is usually such that the preponderant direction of the energy flow in the bond is positive, e.g. from the drive to the load. It can be called natural direction of energy flow.

Nevertheless, the real energy flow can temporarily reverse its direction, e.g. if a machine with an appreciable moment of inertia (flywheel) is braked by its drive. Generally, during the discharge of I- and C-elements the direction of energy flow reverses its sign. As an example from mechanics, when a flywheel is braked the torque reverses and the rotation frequency becomes progressively smaller. With a spring (C-element) the velocity changes its sign, whilst the direction of force remains equal and the force is reduced progressively. These processes so occur that the stored energy is precisely recuperated.

The natural direction of energy flux for one-ports is out from the source and into the R-element. For I- and C-elements one usually chooses as positive the energy flow into the element, as shown in Fig. 2.4. For two-ports, transformers, and gyrators there is no preferred choice of positive energy flow directions.

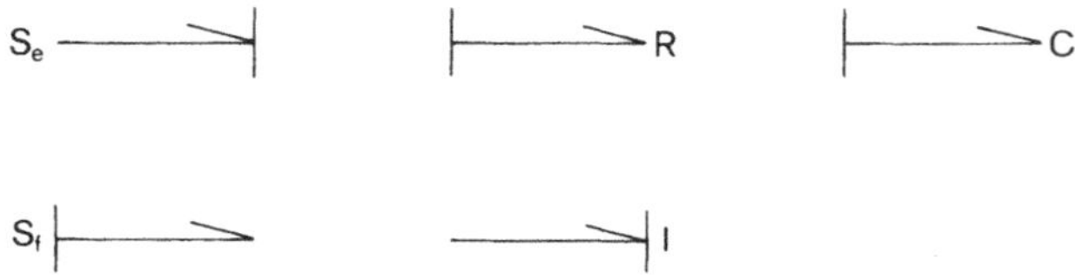

Fig. 2.4. Different one-ports with causality and positive directions of energy flux.

Fig. 2.5. p-Junction with different directions of energy fluxes.

Referring to the signs at the junctions, all energy fluxes add up to zero if their directions are taken positive towards the junction whence the half-arrow points toward it (Fig. 2.5, left). This means

$$\dot{E}_1 + \dot{E}_2 + \dot{E}_3 = 0 \qquad e_1 f_1 + e_2 f_2 + e_3 f_3 = 0 \tag{2-1}$$

or, since with a p-junction all efforts are equal,

$$f_1 + f_2 + f_3 = 0 \tag{2-2}$$

With this choice of signs the sum of positive flows is equal to zero.

On Fig. 2.5, right, the inflow over bond (1) and the outflow over bonds (2) and (3) must be taken positive. Therefore

$$\dot{E}_1 - \dot{E}_2 - \dot{E}_3 = 0 \qquad e_1 f_1 - e_2 f_2 - e_3 f_3 = 0 \tag{2-3}$$

and, due to the equal efforts at the p-junctions,

$$f_1 - f_2 - f_3 = 0 \quad \text{or} \quad f_2 = f_1 - f_3 \tag{2-4}$$

Equation (2-4) contains the usual sign convention of block diagrams, where the error variable f_2 is equal to the difference of the input variable f_1 and the output variable f_3.

2.3.3. *Rules for the selection of causality*

Causality is not a natural fact but selected by the engineer to facilitate computation. The choice is free in principle but subject to the following limitations:

(1) Sources always impose an effort or a flow to the system. Consequently the causality must be selected as shown in Fig. 2.4 by the cross-strokes. If one were, for instance, to select for the voltage source a different causality where the system imposes an effort on the source, then we have very large (in principle infinite) flows if both voltages do not coincide.

As example one could try applying a voltage of 150 V to an electrical household outlet of 220 V. This leads to excessive current and damage unless a fuse interrupts the outlet.

(2) With linear I-elements the flow is proportional to the time integral of effort. The direction f from e implies thus an

integration and is called integral causality. Similarly with a C-element the effort is the integral of the flow and the direction e from f gives an integration or integral causality. Both integral causalities are indicated on the I- and C-elements of Fig. 2.4 by cross-strokes.

(3) Integral causality is much more useful for machine computation as analogue computers can only carry out integrations and no derivations. Much of the art of programming analogue computers is to change the diagram or the equations such that only integral causalities exist. This can be done very conveniently on a bond graph. Furthermore integral causality is almost invariably preferred for digital computers.

(4) In certain laws, for instance where the pressure drop is approximately equal to the square of flow, the function is single valued only with one causality – in this example with the volume flow as input and the pressure as output variable. Similarly with dry (Coulomb) friction the force is a single-valued function of velocity but not vice versa. Generally the causality must be chosen such that the elements are described by a single-valued function – in the above example in the direction pressure from volume flow as indicated on the R-element of Fig. 2.3. With linear R-elements, on the other hand, there is no preferred causality.

There exists a curious reversion in mathematics by the fact that with mathematical formulae the derivation can always be carried out, whilst integration is much more difficult and analytically feasible only with simple formulae. Consequently, many laws of physics are taught in derivative causality, e.g. the induction law of electronics mentioned above. On the other hand, numerical computation, especially on analogue and digital computers, integration can always be carried out (as sum of many parts), whilst the derivation brings some difficulty. The student of mathematics is confronted with this fact for the first time when he learns that with convergent power series term-by-term integration always improves convergence, whilst derivation makes the convergence worse and might render the series divergent indeed.

The above rules of causality lead to a contradiction in some engineering systems, called a causality conflict. Such conflicts oblige the engineer to occasionally admit derivative causality.

Referring to the causality of two-ports, a transformer with an effort input from the left acts with another effort according to the transformer ratio to the right. The load reacts from the right with the flow which the transformer returns to the left (Fig. 2.6, top left). Alternatively, the effort can act from right to left and the flow from left to right as shown on Fig. 2.6 (bottom left).

Fig. 2.6. Different causalities on transformers and gyrators.

If an effort acts on a gyrator from the left, it determines the flow to the right depending on the gyrator ratio, and the effort reaction from the right determines the flow at left (Fig. 2.6, above right). The other possibility is that the flow left determines the effort at right and the flow reaction from the right, the effort at left, as shown on Fig. 2.6, bottom right.

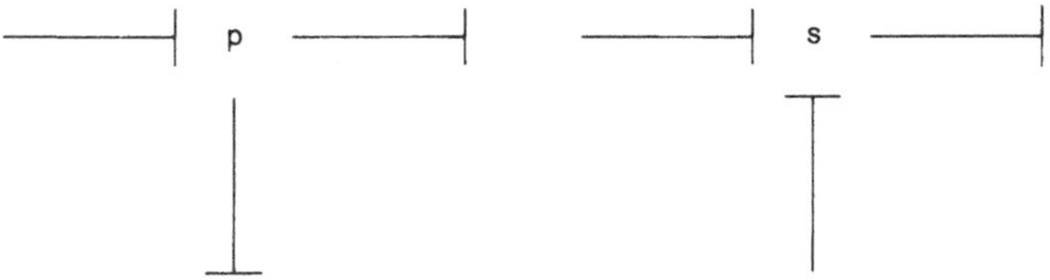

Fig. 2.7. Causalities on p- and s-junctions.

Referring now to junctions, an input effort determines all other efforts in a p-junction. This leads to the causality of Fig. 2.7 at left. The s-junctions contain a common flow, hence one input flow determines all output flows, whilst one effort is determined as sum or difference of all input flows as shown on Fig. 2.7 (right). The bonds of Fig. 2.7 can be arranged differently, but the following always holds:

(1) With p-junctions there are:
- one input and several output efforts;
- several input flows and one output flow

(2) s-Junctions contain:
- several input efforts and one output effort;
- one input flow and several output flows.

It should be noted that the words input and output refer to the direction of causes and effects or causality and not to the direction of energy flux, which is indicated separately (§ 2.3.2).

2.4. Transformation and Simplification of Bond Graphs

At this point several transformations of bond graphs will be given which allow substantial simplification of the bond graphs for some mechanical and hydraulic systems.

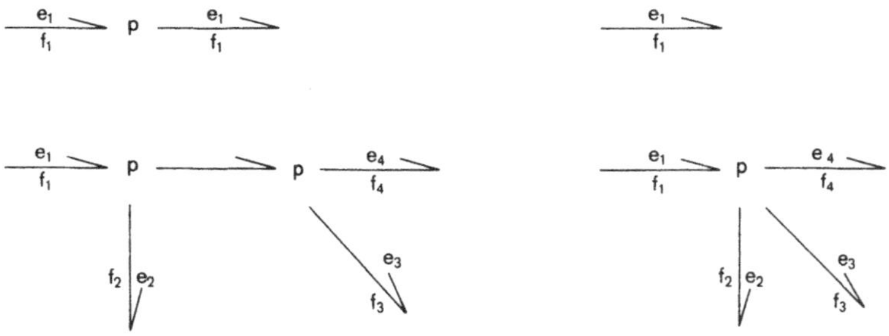

Fig. 2.8. Simplification of junctions.

Referring to Fig. 2.8 (top), if a junction has only two bonds and the positive energy flows are chosen as shown, then both efforts and both flows are equal, and the junction can be replaced by a single bond. This applies both for p- and s-junctions.

If two junctions are following one another, as shown in Fig. 2.8 (bottom), they can be combined to a single junction. This is because in both cases the efforts are equal and the flows related by

$$f_1 - f_2 - f_3 - f_4 = 0 \qquad (2\text{-}5)$$

as seen easily by eliminating the flow in the bond between both p-junctions. A similar relation holds for s-junctions only with effort and flow interchanged, so that two consecutive s-junctions can be combined in the same way.

The position of s- and p-junctions cannot be interchanged (the junctions do not comute), but the following important transformation applies.

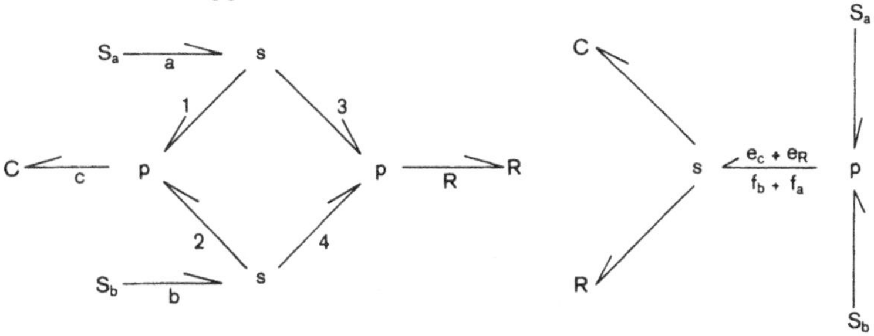

Fig. 2.9. Simplification of a ring of p- and s-junctions.

Figure 2.9 (left) is a bond graph with a loop of two p- and two s-junctions as frequently encountered in mechanics. The s-junctions lead to other bonds a and b connected with the sources S_a and S_b. On the p-junction there are the bonds C and R with the one-ports shown as example. It is now possible to replace the loop by the two junctions of Fig. 2.9 (right) under the condition that the signs on both p-junctions are chosen equal. In other words both f_c and f_r must be the sum of f_1 and f_2 or f_3 and f_4 respectively. The flows f_c and f_r can also be the difference of these flows, but not one of them the sum and the other one the difference.

The flow in the horizontal bond of Fig. 2.9 (right) is equal to the sum of the flows of the sources since both s-junctions of the original figure assure the equality of the flows. The effort in this bond is the sum of the efforts on the bonds C and R. A mathematical proof of this equivalent is obtained by calculating the flows and efforts from the equations of the different junctions, whence both parts of Fig. 2.9 lead to the same relations.

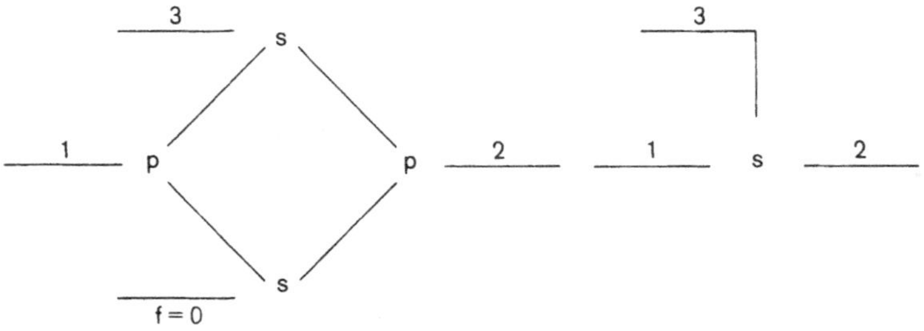

Fig. 2.10. Further simplification of Fig. 2.9 if the flow in the lower junction vanishes.

Important simplifications of the bond graph loop result if the flow in one s-junction vanishes as shown on Fig. 2.10. Then the adjacent bonds disappear and the p-junctions in the centre retain only two bonds. They can then be eliminated as above, and all that remains is the simple s-junction of Fig. 2.10 (right).

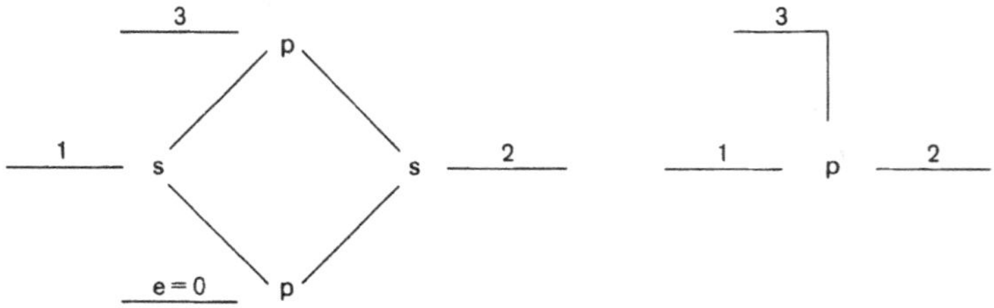

Fig. 2.11. Simplification with vanishing effort in the lower p-junction.

Similarly, if in a p-junction of the loop one effort is equal to zero (Fig. 2.11), it vanishes together with the adjacent bonds. Consequently only the simple p-junction of Fig. 2.11 (right) remains. Simplifications of this kind arise in practice by setting the effort (the potential in electrical systems) of one p-junction equal to zero, which means that all other potentials are measured from the value of this potential.

A transformer can be interchanged with a p-junction or a s-junction as shown in Fig. 2.12 if the other bond receives a new transformer with the reciprocal ratio l/i. This is intuitively evident but follows also from a mathematical manipulation of the relevant equations.

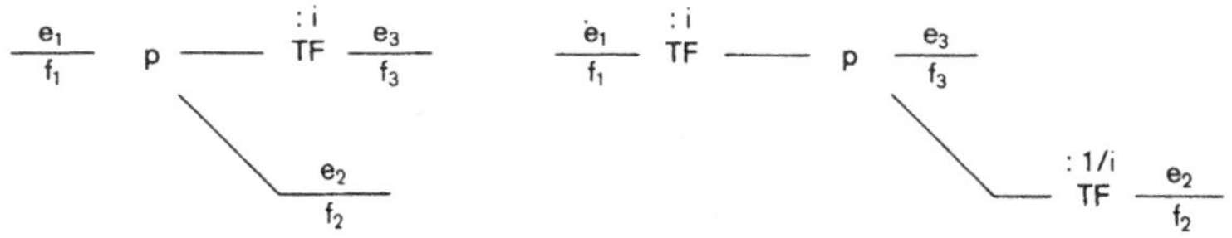

Fig. 2.12. Interchange of a transformer with a p-junction.

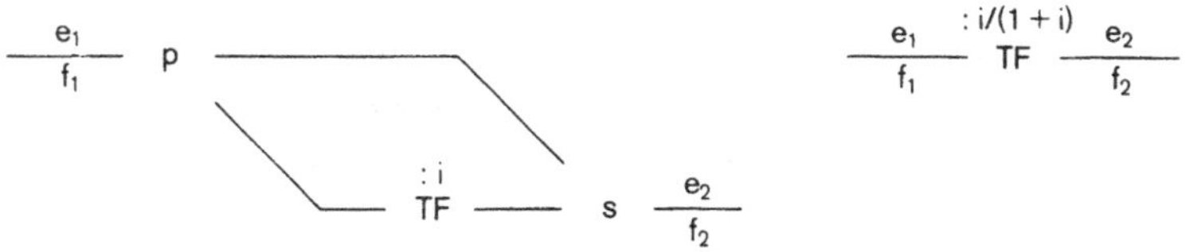

Fig. 2.13. Simplification by a transformer with different transforming ratio.

Finally, according to Fig. 2.13 a transformer with the ratio i together with a direct bond between a p- and s-junction can be replaced by a transformer alone with the transforming ratio i/(l+i). This transformation is needed for planetary gear reducers (§ 3.3).

2.5. Additional Topics

2.5.1 *Resistance, capacity, and inertia fields*

Table 2.5 contains several multiports or fields which are extensions of the simple R-, C-, and I-elements:

(1) The resistance field or R-field represents an arbitrary, linear, or non-linear relation between flows and efforts in each bond. The bonds can be coupled in such a way that, for

example, a change of the flow in the first bond changes the effort in the second bond even if the external load of the latter remains constant. Without this cross-coupling between the bonds, the R-field would behave like separate R-elements. Example of R-fields are an arrangement of electrical resistors or hydraulic resistors. Although an R-field always dissipates energy in the whole, power can leave through one bond if more power enters the R-field through the other bonds. Thus thère always remains a net energy dissipation or production thermal energy and entropy (Chapter 6).

(2) The capacity field or C-field, where each effort is an arbitrary function of all displacements (time integral of flows) and conversely each displacement, is an arbitrary function of all efforts. The C-field is storing energy temporarily but conserves it in the average. Due to the cross-coupling, energy can be stored through the one bond and taken out through another bond. This is the basis of the use of C-fields as signal transducers with energy storage (§ § 5.3 and 6.3). Similarly, as with R-fields, without the cross-coupling a C-field is equivalent to separate C-elements, storing energy and conserving it on average each by itself.

(3) The inertia field or I-field has an arbitrary relation between the flows and the momenta (time integrals of efforts) and also conserves energy on average. Examples are a network of electrical inductors and a rigid beam with mass and moment of inertia.

Table 2.5 does not contain the following more complex fields necessary for the explanation of the physical operation of some components:

(1) The IC-field, a mixture of I- and C-fields, where at one bond (or several) there is a functional relation between momentum and flow and another bond (or several) a relation between displacement and effort, with possible coupling between all bonds. An example with only two bonds is a magnetic actuator (§ 5.3.3).

Table 2.5. *Standard elements of bond graphs: fields*

Symbol	Designation	Examples		
		Electrotechnics	Fluid power	Mechanics
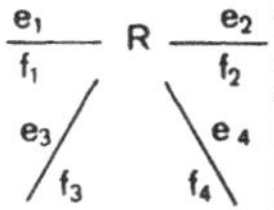	R-field $e_i = F_{Ri}(f_1, f_2 \ldots f_n)$ $f_i = F_{Ri}^{-1}(e_1, e_2 \ldots e_n)$	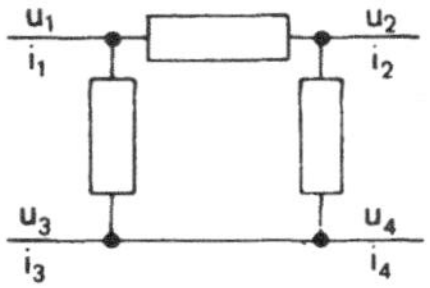 Network of elec. resistors	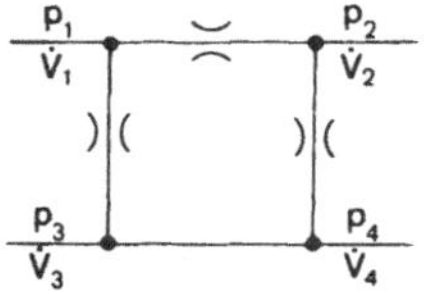 Network of hydraulic resistors	
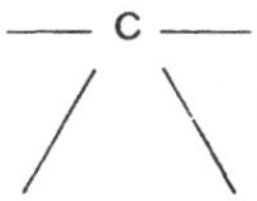	C-field $e_i = F_{ci}(q_1, \ldots q_n)$ $q_1 = F_{ci}^{-1}(e_1, \ldots e_n)$	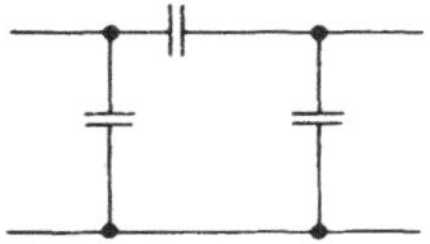 Network of elec. capacitors	–	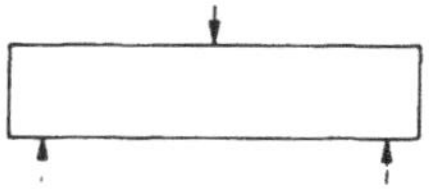 Deformable beam without mass
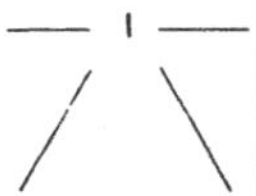	I-field $p_i = F_{Ii}(q_1, \ldots f_n)$ $f_i = F_{Ii}^{-1}(e_1, \ldots e_n)$	 Network of inductors	–	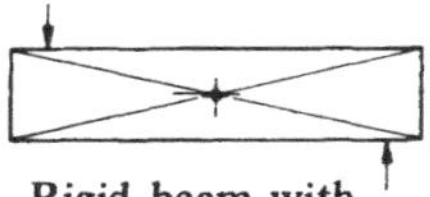 Rigid beam with mass and moment of inertia

(2) The RS-field is a combination of an R-element and a source delivering the power dissipated by the R-element. This field is used for thermal effects, where the source produces an entropy flow at an absolute temperature normally fixed by the environment (cooling) (§ 6.3). More complex RS-fields appear in § 8.3, obtained from an R-field with several bonds by adding a source for the entropy flow corresponding to the dissipated power.

It should be mentioned at this point that in linear fields the general functional relation of the variables is described by matrices. The different efforts and the different flows in R-fields can be interpreted as the components of a vector. Thus the effort vector is obtained from the flow vector by multiplication with the matrix, and similarly the flow vector from the effort vector by multiplication with the inverse matrix. This is a simplification (linerization) of the basic equations of R-fields on Table 2.5. Similarly, in a linear I-field the momentum vector is connected with the flow vector by a matrix and, finally, in a C-field the displacement vector with the effort vector.

Whilst the question of the different possible causalities in fields is treated in ref. 2, it should be noted here that the two different forms of the equations of each field in Table 2.5 correspond to different causalities.

The fields may appear artificial at first sight but they provide a solution for many problems, not only in complex engineering systems but also of relatively simple components. As an example, an electrical condenser with movable plate is a C-field with two bonds (§ 5.3), where one bond supplies electrical energy and the other bond mechanical energy with the force and displacement of the plates. An example from thermodynamics is a gas, where pressure and volume flow travels through the mechanical bond and absolute temperature and entropy flow through the thermal bond (Chapter 6).

Referring to the nature of the energy stored in C- and I-fields, it can enter either through one or the other bond. But once energy is inside the field it is independent from the supply route and depends only on the displacements and efforts (in C-fields). In other words the stored energy in the example of the electric capacitor with movable plates is not of the electrical or mechanical kind but simply universal energy. This throws an interesting light on the concept of energy in physics and engineering science.

2.5.2. *Tetragon of state variables*

The four variables – effort, flow, (generalised) momentum, and displacement, are also called state variables since their values determine the state of an engineering system. These variables can be displayed in a tetragon after Paynter to illustrate their relation (Fig. 2.14). The figure also indicates the action of the several one-ports, in

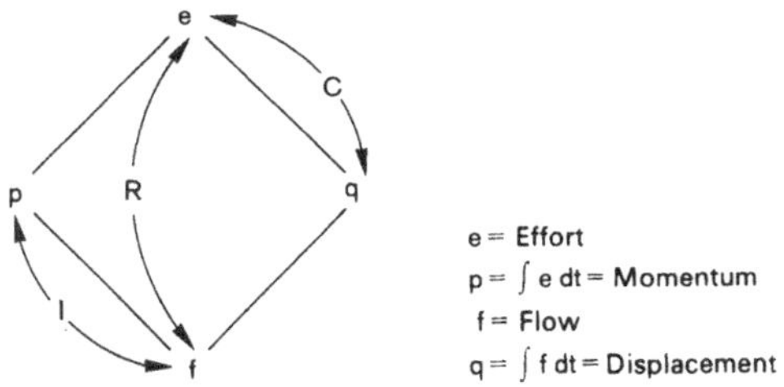

Fig. 2.14. Paynter's tetragon of the state variables, effort e, momentum p, flow f, and displacement q.

particular that an R-element gives a relation between effort and flow, an I-element between momentum and flow, and C-element between effort and displacement.

2.5.3. *Non-linear impedances*

Non-linear impedances are described in general by their characteristics between flow and effort as shown on Fig. 2.15 for a resistor (restrictor in fluid power) with a quadratic relation between pressure drop and volume flow. The curve can be approximated in an operating point by its tangent. The inclination of the tangent gives the ratio between the increments of pressure and volume which is defined as the tangent impedance Z_{tg}. Furthermore, the ratio of pressure drop and volume flow at the operating point is defined as the secant impedance Z_{sec}, since it corresponds to the inclination of the secant between the operating point and the origin.

$$Z_{tg} = \frac{p - p_r}{\dot{V} - \dot{V}_r} \qquad Z_{sc} = \frac{p_r}{\dot{V}_r} . \tag{2-6}$$

The reciprocals of these impedances are the tangent and secant admittances.

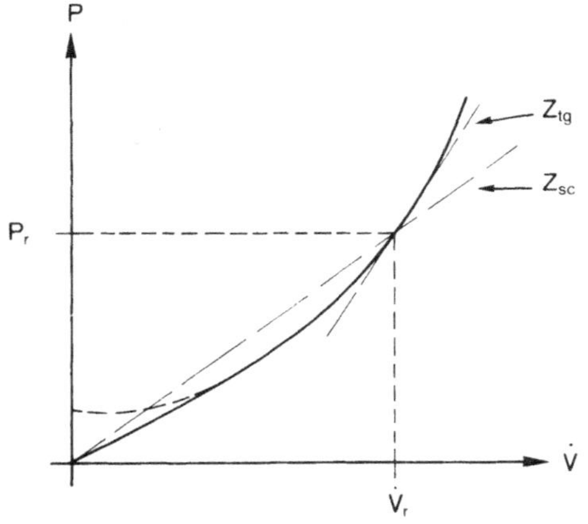

Fig. 2.15. Non-linear characteristic with tangent and secant impedances.

With linear characteristics starting from the origin, both impedances become equal since tangents and secants are identical and the ratio of tangent and secant impedances is equal to 1. The further this ratio deviates from 1 the more non-linear is the characteristic. In other words, this ratio is a quantitative indication of the non-linearity of a characteristic.

The distinction between tangent and secant impedances is outside the scope of bond graphs, but is important for the design and dimensioning of technical systems. Secant impedances in particular give the relation between the maximum values of the variables for which the components must be designed.

CHAPTER 3

Applications from Mechanical Engineering and Design

3.1. Simple Installations

3.1.1. *Introduction*

This chapter shows bond graphs in mechanical systems and gives an introduction to their methodical application. A machine is a complex system according to its detail drawings, but usually some effects are preponderant and can be well modelled by bond graphs. This will be shown in several examples below, including selection of the positive energy directions and the causalities. The next step is to augment the bond graphs in the sense of Chapter 2, leading to the block diagrams with resonances and damping.

Figure 3.1 (left) shows a machine supported from an immobile base by springs and friction elements. It is excited by a periodical force source, e.g. by an unbalanced rotor, and we would like to determine the resulting movement or oscillations of the machine. Only vertical movements are admitted, supposing that suitable guiding members prevent rotations and transversal movements.

Figure 3.1 (centre) contains a bond graph with words for excitations, machine, suspension, and base with simple lines for the power bond with energy exchange. In order to obtain the bond graph with standard elements (right), the following idealizations are introduced:

(1) The base is so large and massive that it does not move. No power is transmitted on it and consequently the bonds between suspension and base vanish.

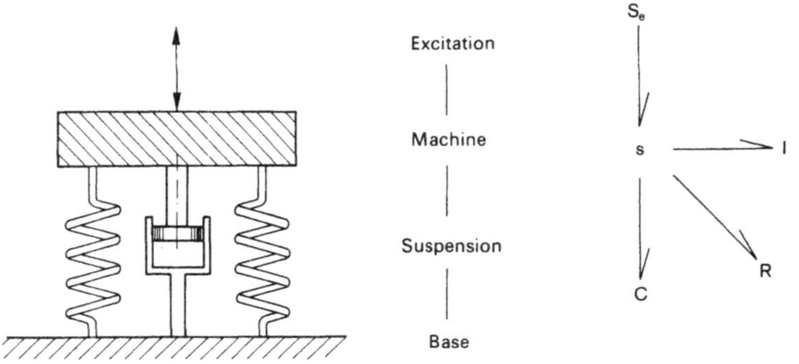

Fig. 3.1. Machine with spring and friction elements on base, schematic (left), word bond graph (centre), and bond graph with one-ports (right).

(2) The excitation is a periodic force source supplying a force independent of velocity.
(3) The machine itself is rigid with negligible internal deformations and moves like a mass point with the mass m as shown by the I-element.
(4) The suspension consists of (a) a spring with the spring constant k as C-element, and (b) a friction member as an R-element.

Between the elements of Fig. 3.1 there is an s-junction as the velocities are equal and the different forces are added. This completes the simple bond graph which can be augmented by choice of the causalities and of the positive energy fluxes. Spring, friction member, and mass action are one-ports as described in § 2.2.

Whilst until this point the C- and R-elements could be non-linear, we shall now suppose that these one-ports are linear, where the spring force is proportional to the displacement and the friction force proportional to the velocity. The force of I-elements is always proportional to the acceleration if one excludes effects of relativity after Einstein, i.e. if the velocity is well below the velocity of light.

Figures 3.2 and 3.3 show bond graphs augmented from Fig. 3.1

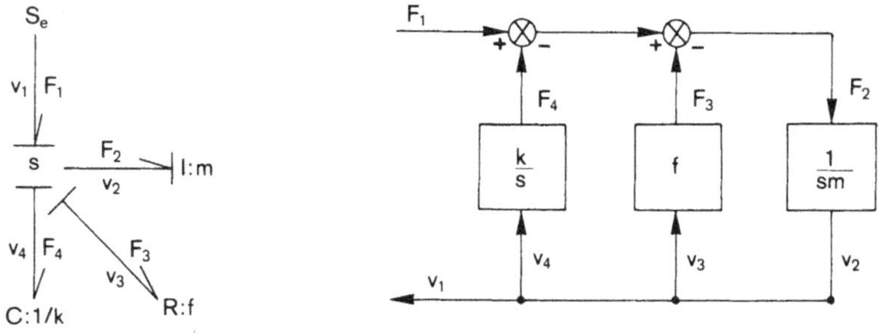

Fig. 3.2. Augmented bond graph and block diagram of the machine of Fig. 3.1.

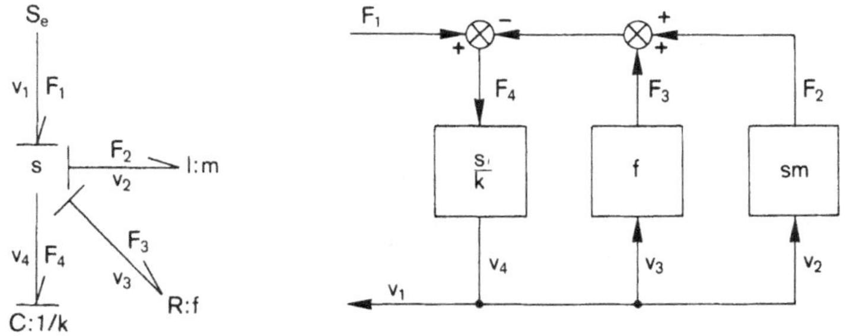

Fig. 3.3. Augmented bond graph for Fig. 3.1. in different causality.

with half-arrows for the positive energy directions and in two different causalities. Furthermore, the bonds are numbered 1–4 and consequently the velocities denoted v_1–v_4 and forces denoted F_1–F_4. As usual the direction of energy fluxes are chosen such that they are positive when power leaves the sources and enters the one-ports. The effort source imposes an effort to the system and determines the causality with the effort acting on the s-junction (Chapter 2). This is indicated by the cross-stroke at the bond between the effort source and the s-junction.

The further causalities of Fig. 3.2 are such that the force of the effort source acts on the mass which responds with the velocity v_2, but after deducting the spring and friction force (for the signs see below). Both are functions of the velocity.

The bond graph of Fig. 3.2 (left) can be interpreted as representation of a number of equations obtained from Table 2.2 for the one-ports and for the junctions. The bond graph contains also the causalities, which here are all integral since there are only integrations and proportional relations but no derivations with respect to time.

Figure 3.2 (right) contains the corresponding block diagram with the spring constant k, friction constant f, and mass m, where the relations for the one-ports are written into the blocks. The block diagram is obtained by inserting the equation for the one-port by representing the equality of the flows (torque) by the mixing points. In order to be clearer the bond graph is arranged such that the upper connections contain only efforts and the lower connections only flows.† The variable s is a derivation operator appearing here twice in the denominator and indicating an integration with respect to time corresponding to the integral causality.‡

Both the bond graph and the block diagram of Fig. 3.2 contain the same information. This is also clear from the fact that all forces and velocities (efforts and flows) appear in both representations. The difference is that each connection of the block diagram contains only one variable and that to a bond of a bond graph there are always two corresponding connections for effort and flow.

Figure 3.2 is an example of how an s-junction becomes a branch point for the flows and a mixing point for the efforts in a block diagram. It will be seen later on that a p-junction turns in to a branch point for the efforts and into a mixing point for the flows.

Regarding the signs it should be noted that with the half-arrows of Fig. 3.2 the energy fluxes at the s-junctions are as follows:

$$\dot{E}_1 - \dot{E}_2 - \dot{E}_3 - \dot{E}_4 = 0; \quad F_1 - F_2 - F_3 - F_4 = 0$$

†This arrangement was introduced originally by H. Paynter at MIT and can be considered as a preliminary step towards bond graphs.

‡Multiplication with s implies derivation, division by s integration with respect to time, the Laplace transform s becomes the complex signal frequency $\sigma + i\omega$.

where the second form is obtained from the fact that E = vF and that all velocities are equal on s-junctions. Consequently the following is obtained:

$$F_2 = F_1 - F_3 - F_4$$

as represented by the signs at the mixing points of Figs. 3.2 and 3.3.

Figure 3.3 is distinguished from Fig. 3.2 only by the different causalities, since here the exciting force acts on the spring (C-element) and produces the velocity which by itself determines the other forces that have to be deducted from the exciting force. This leads to two derivations, namely between spring force and velocity and between velocity and accelerating force. Consequently, Fig. 3.3 is a bond graph in derivative causality. The transfer function between velocity of the machine and the force is obtained from the block diagram by the formula of Mason, where Figs. 3.2 and 3.3 give the same result in accordance with the inversion theorem. In Fig. 3.3 both blocks on the right can be contracted into a block with the gain f + ms and set equal to G of (1-1) since they are in the feedback connection between v and F. The gain of the block on the left becomes F of (1-1) and we obtain for the required transfer function by using (1-2):

$$\frac{v_1}{F} = \frac{s/k}{1 + s\,(f + ms)/k} = \frac{s/k}{1 + 2\zeta\, s/\omega_r + s^2/\omega_r^2} \tag{3-1}$$

with

$$\omega_r^2 = k/m \qquad 2\zeta = \omega_r\, f/k$$

The second form of eqn. (3-1) is the so-called canonical form of a second-order system having a resonance with a frequency ω_r and a damping solution ζ, which should have a value of about 0.2 for satisfactory operation. The control engineering literature contains the further treatment and interpretation of eqn. (3-1).

The above detailed treatment of the bond graph for simple oscillating systems allows the setting of block diagrams systematically in different causalities and the determination of the signs in a clear manner. The bond graph contains, as mentioned already, the same information in more compact form, and is therefore useful especially for more complex systems where block diagrams with many blocks and feedback connections become impracticable.

3.1.2. *Torsional oscillations through a gear-reducer*

Figure 3.4 shows, as a further example, the drive of a pump through a compliant shaft and a gear-reducer as a schematic (above)

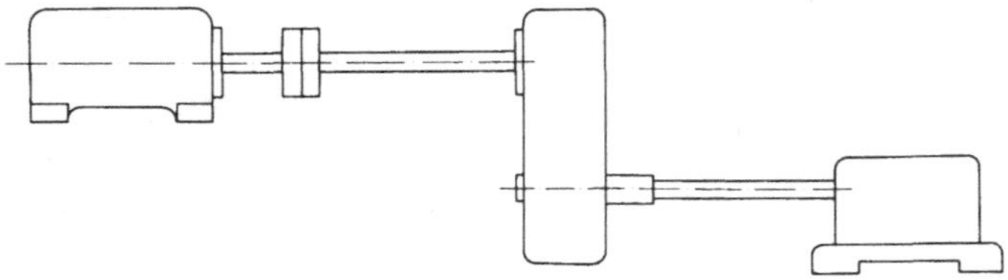

Drive ——— Shaft ——— Gear-reducer ——— Pump

Fig. 3.4. Pump drive through a gear-reducer, schematic (top) and word bond graph (bottom).

and as a word bond graph (below). In the following bond graph, Fig. 3.5, the drive is taken as a controlled flow source where the rotation frequency is independent of torque, as approximately true with electric motors with constant field current (shunt-excited machine, § 5.1). The shaft has the compliance k, but no moment of inertia, and is therefore a C-element; the gear-reducer is an ideal transformer with the reduction (transformation) ratio i but without losses or moment of inertia. The pump has a moment of inertia J and the torque absorbed by the pumping action is represented by an R-element running with the same rotation frequency thus behind an s-junction.

The compliant shaft is connected through a p-junction, since the torques are equal and the torsion angle (angle of twist) of the shaft is equal to the difference of the rotation angles of the drive and the

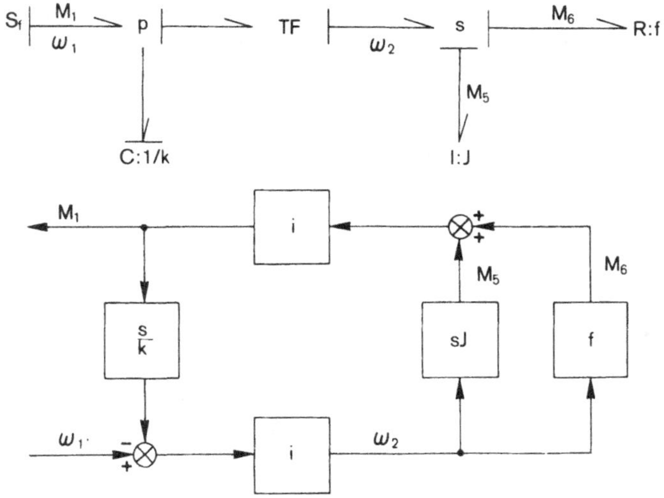

Fig. 3.5. Bond graph (top) and block diagram (bottom) from Fig. 3.4.

gear-reducer. In addition, Fig. 3.5 contains positive energy directions and a possible selection of the causalities.

The lower part of Fig. 3.5 shows the corresponding block diagram. It should be noted that gear-reducer turns into two blocks with the gain i, the first one to indicate the reduction of the rotation frequency and the second one to indicate the increase of torque.

The block diagram in Fig. 3.5 is relatively simple but contains two derivative causalities. It corresponds to the description that from the input rotation frequency ω_1 the rotation frequency of the shaft twist has to be deducted to obtain the rotation frequency ω_2 of the pump after the speed reduction. The latter determines the accelerating torque M_6 which, after a further reduction i, produces the torque M_1 responsible for the twist of the shaft.

Figure 3.6 contains a bond graph and a block diagram in different, integral causalities, where the gains of the blocks and the effect of the transformer are the reciprocal of Fig. 3.5. This is due to the fact that the signal runs through the blocks in the inverse direction, similarly as after inversion of a simple block diagram.

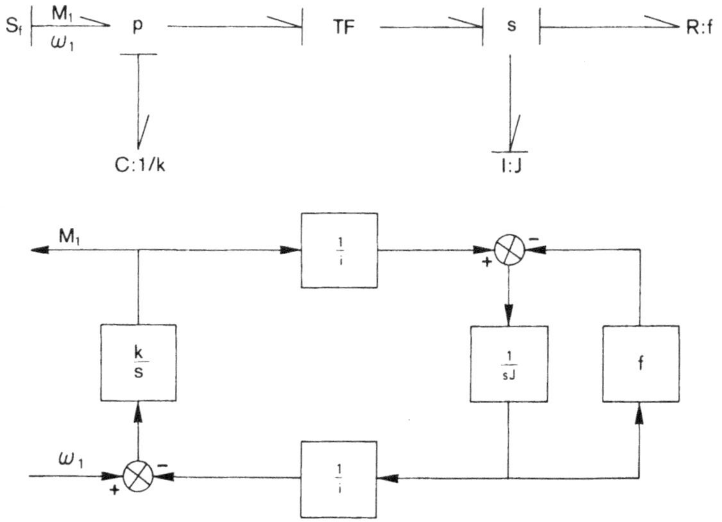

Fig. 3.6. Bond graph (top) and block diagram (bottom) from Fig. 3.4.

The block diagram is now more complex since it contains an internal feedback loop. The operating of the pump drive according to Fig. 3.6 in words is such that from the input rotation frequency the frequency of the gear-reducer must be deducted to obtain the twist of the shaft. This produces the torque of the gear-reducer, the pumping torque must be deducted (mixing-point top-right) in order to obtain the rotation frequency by the moment of inertia. This rotation frequency is responsible for the pumping torque and for the feedback rotation frequency through the gear-reducer.

It is seen that the verbal description of the operation of Fig. 3.6 is considerably more involved, and consequently the causality of Fig. 3.5 is preferably for explanations and study. However, for automatic computation the integral causalities of Fig. 3.6 are necessary.

Block diagrams are simplified if the causalities are selected such that the s-junctions have a flow as input and the effort as return action (Fig. 3.5). With p-junctions the efforts should act on the junction which reacts with a flow. Observing this rule produces simple block diagrams with few internal feedback loops but in the above drive it is in contradiction with the desired integral causalities and thus an example of a causality conflict (§ 2.3).

In order to obtain the dynamic relation between rotation frequency and torque it is convenient to contract in Fig. 3.5 both blocks at right to a block with the gain sJ + f. Consequently the gains F and G of (1-1) become

$$F = i^2 \, (sJ + f) \qquad G = s/k$$

and thus the change of input torque due to a change of the rotation frequency

$$\frac{M_1}{\omega} = \frac{i^2 (sJ + f)}{1 + (sJ + f)\, i^2\, s/k} = \frac{i^2 f(1 + s/2\zeta\omega_r)}{1 + 2\zeta\, s/\omega_r + s^2/\omega_r^2} \tag{3-2}$$

with

$$\omega_r^2 = \frac{k}{i^2 J},$$

$$2\zeta = \frac{\omega_r f i^2}{k} = \frac{i^2 f}{\sqrt{kJ}}.$$

Here we have a resonance through the transformer or gear-reducer. Hence the moment of inertia in the formula of the resonance frequency is multiplied by the factor i^2 as well as the friction constant f for the damping. Furthermore, due to its structure (3-2) contains a term with $s/2\zeta\omega_r$. It produces a derivative action at high signal frequency as known from control engineering.† The gear reducer is thus an impedance transformer between two bonds of the same energy domain and transforms the impedance by a factor i^2.

Resonances through transformers are frequent in combined electrical, hydraulic, or mechanical systems, where the transformer can also be a class transformer between two different energy domains as described in § 2.2.

The above example concludes the simple mechanical applications described in detail including the determination of the block diagram and the transfer function.

†The signal frequency should not be confused with the rotation frequency. Here it is equal to the change of rotation frequency with time.

3.1.3. *Combined one-ports and transformers*

One-ports are occasionally connected to the system through a transformer as shown on the example of a linear R-element in Fig. 3.7 as a bond graph (top left) and as a block diagram (bottom left).

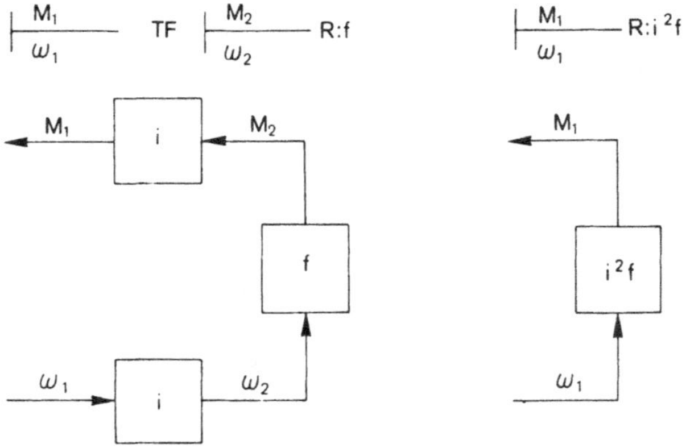

Fig. 3.7. Substitution of a transformer and an R-element by an R-element multiplied with i^2 (i = transformer ratio).

The input variable, here a rotation frequency, is initially increased by the transformer ratio i before it acts on the R-element. The resulting effort, here the torque, is again multiplied by i. The impedance of the combined one-port and transformer is consequently increased by the factor i^2 as shown in Fig. 3.7 (right) as a bond graph and as a block diagram. Similar relations hold for I- and C-elements not only in mechanics but also in the other engineering disciplines. Especially in electronics the change of the values of R-, I-, and C-elements with the square of the transformer ratio i is well known. It is universally valid for efforts and flows and is clearly displayed by bond graphs.

With non-linear R-elements we set

$$M_2 = F(\omega_2) \qquad M_1 = i\,M_2 \qquad \omega_2 = i\,\omega_1$$

from which we obtain

$$M_2 = i\,F\,(i\,\omega_1)\cdot \tag{3-3}$$

The argument of a non-linear function is multiplied by i and the output variable (torque) again by i, which reduces in the linear case to the multiplication by i^2 as mentioned. Similar relations hold for non-linear I- and C-elements. With different (inverse) causalities the value of one-ports is divided by i^2 by the transformer, i.e. instead of $1/f$ one obtains $1/i^2 f$.

3.2. Bond Graphs of Mechanical Drives

3.2.1. *Complex mechanical system*

For an example of a complex mechanical system we refer again to the pump drive of Fig. 3.4. Here we shall further suppose that both drives, reduction gear, and pump have an appreciable moment of inertia, that both shafts between these components have significant rotational compliances, and that the gear-reducer has a loss torque. Figure 3.8 shows the corresponding bond graph in two different causalities.

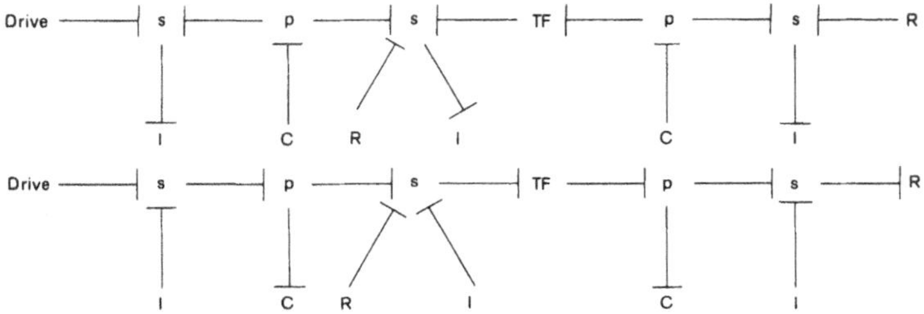

Fig. 3.8. Bond graph of a complex mechanical system, top and bottom in different causalities.

In Fig. 3.8 the drive is represented by a word, but an I-element for its moment of inertia has been added through an s-junction. The following C-element behind a p-junction (equal torque, addition of rotation angles) represents the rotational compliances of the shaft between drive and gear-reducer. The adjacent R- and I-elements indicate the loss torque and moment of inertia, whilst the reducer itself acts as a transformer. The following C-element at the p-junction indicates the compliance of the shaft between gear-reducer and pump, which itself is modelled by the I-element for its moment of inertia and by the R-element for its pumping moment.

Figure 3.8 contains only integral causalities since in all C-elements the flow is the input and the effort (torque) the output, whilst on the I-elements the effort is input and the flow output. Consequently it is particularly useful for automatic computation or electronic data processing (EDP). Therefore all bonds must be numbered consecutively (as in Fig. 3.5), and then the nature and numerical values of the different elements programmed into the computer as described in Chapter 7.

The causality of Fig. 3.8 (bottom) is more useful for building a block diagram and for a general description of operation. It contains less internal feedback loops since here the efforts act on the p-junction. The disadvantage is that both C-elements and the I-element at the centre right give derivative causalities.

The R- and I-elements for the friction and the moment of inertia of the gear-reducer can be connected also to the other side, i.e. to the right of the transformer. The value of these elements would then have to be multiplied by i^2 as shown in § 3.1.3.

The bond graphs of Fig. 3.8 can be converted into block diagrams in a straightforward manner, but they become somewhat involved and will therefore not be represented here. As mentioned, block diagrams carry the same information as fully augmented bond graphs. With some experience one can recognize in Fig. 3.8 several possibilities of resonance, which will be determined by machine computation in practice.

Machine computation is possible only when the numerical values of all elements are available. This applies not only to bond graphs but also to block diagrams and mathematical formulae, which give quantitative results only with the numerical values of all constants.

3.2.2. *Characteristics of the drive*

The characteristics of the drive determine how the word "drive" of Fig. 3.8 has to be replaced by bond-graph elements. A characteristic in this sense describes the relation between torque and rotation frequency which frequently depends on a control signal.

Only linear characteristics allow a completely satisfactory replacement by standard bond-graph elements. Many characteristics, however, can be linearized if small derivations from an operating point are considered (§ 1.5) whence the bonds contain only these deviations.

Figure 3.9 contains three different characteristics a, b, c each depending on a control signal (parameter) as denoted by 1, 2, 3. The

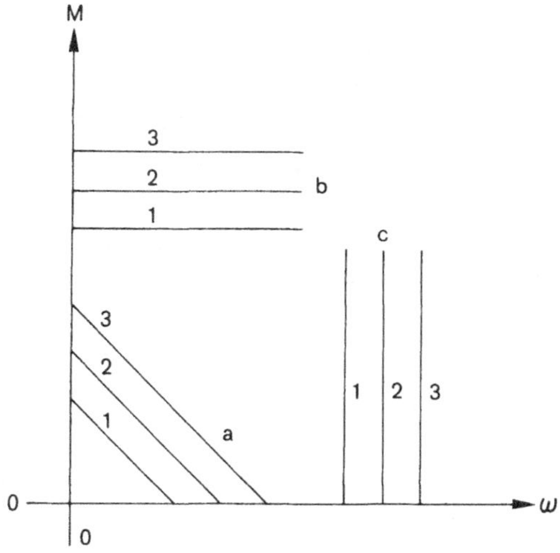

Fig. 3.9. Characteristics of a mechanical drive.

characteristics can be represented by a controlled effort source with R-elements in series as shown on Fig. 3.10 (left) as a bond graph.

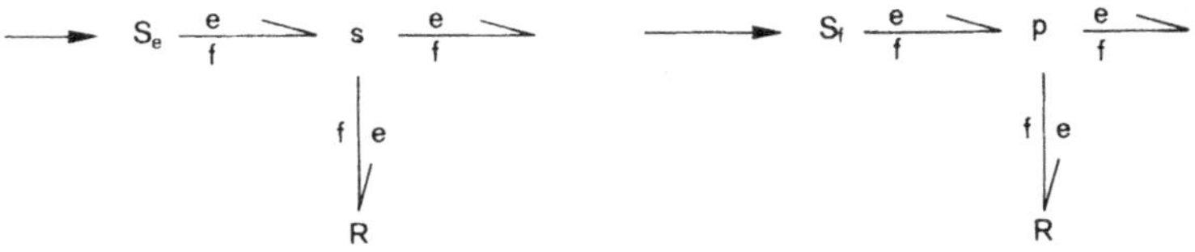

Fig. 3.10. Representation of a characteristic by an effort source with a series resistor (left), and by a flow source with a parallel resistor (right).

The value of the source is equal to the value of the effort variable with the flow zero, i.e. to the intersection of the characteristics with the y-axis. The series resistance models the reduction of the effort with the flow drawn from the system and corresponds to the inclination of the characteristic.

Alternatively, the characteristics can be modelled by a flow source with a parallel resistance (Fig. 3.10 (right)). The value of the source is equal to the flow with vanishing effort, and the parallel resistance indicates the reduction of flow with opposing effort. In both cases the resistance (impedance) has the same value.

The above considerations are applicable both to hydraulic and electronic sources. Especially in electronics the controlled flow source (current source) with parallel resistance appears frequently. The equivalence of the representation by both sources is known as the theorem of Thevenin. This again is a theorem of systems engineering represented very clearly by bond graphs.

With the characteristic b of Fig. 3.9 the effort (here the torque) is independent of the flow (rotation frequency). Consequently the series resistance is very small and can be left out from the bond graph of Fig. 3.10 (left). The drive is in this case a controlled effort source as well approximated, e.g. by Diesel engines, where the torque is a function of the injected fuel quantity and influenced only very little by the rotation frequency. In electronics this is described as a controlled voltage source without internal resistance.

The characteristics c of Fig. 3.9 show, instead, a flow independent of the required torque. The parallel resistance becomes larger and larger until it reaches infinity if the characteristics are vertical. In electronics this is described as a controlled current source with infinitely high internal resistance. In mechanics such a characteristic is obtained if one machine is driven from a large transmission shaft together with many other machines. This was the current industrial practice in the nineteenth century where the transmission shaft was driven by a steam-engine.

The operating limits must be distinguished clearly from the characteristics. These limits are the values that cannot or must not be exceeded by the variables in order to avoid damage. Examples with Diesel engines are the maximum torque of the maximum rotation frequency. Sometimes even small values of the variables are inadmissible, e.g. with Diesel engines where a minimum rotation frequency must be maintained or the engine will stop entirely.

Such operating limits have some similarity with the speed limits on roads which can be easily exceeded but can be dangerous. Therefore they are frequently monitored automatically by speed limiters, electrical fuses, safety valves, or torque-limiting clutches. However, these questions of system engineering lead outside the subject of bond graphs.

In conclusion the word "drive" of Fig. 3.8 can be replaced by an effort or a flow source with an internal resistance as shown on Fig. 3.10 in accordance with the drive characteristic. The internal resistance acts on the system and produces generally an additional damping of possible oscillations.

3.3. Moving Reference Points and Planetary Gear-reducers

3.3.1. *Rotating casings and torque meters*

So far we have supposed implicitly that the casings or frames of the mechanical components are suspended in an immovable base that returns also the torque reactions of the components. We shall now consider the case of moving reference frames, in particular rotating casings and set up bond graphs for planetary gear-reducers as an application.

The electric generator (Fig. 3.11 (left)) with a rotably suspended casing is our first example. The torque of the shaft and the reaction torque are equal, and the effective rotation frequency is equal to the difference of the frequency of shaft and casing. This is expressed by the p-junction of Fig. 3.11 (right). If the electrical power is for the moment of no interest, the generator can be modelled simply by an R-element dissipating mechanical power behind the p-junction. Both the shaft and the casing are connected in general to other components rotating with the same frequency, as indicated by the s-junctions with bonds on Fig. 3.11 (right).

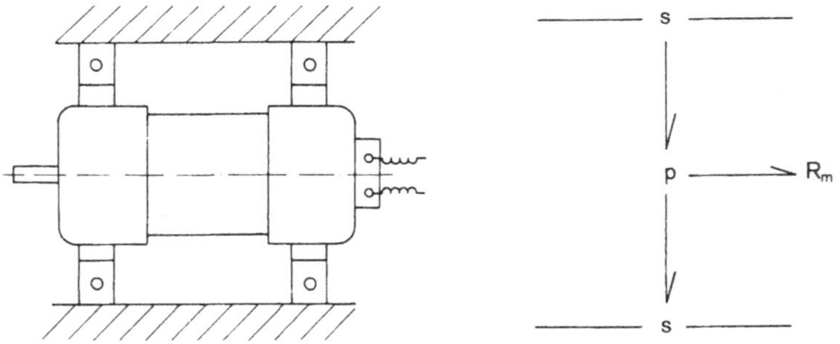

Fig. 3.11. Electric generator with rotable casing, schematic (left) and bond graph (right).

Figure 3.12 shows a bond graph of an electrical torque meter with a d.c. electric machine, where the casing is rotably suspended and connected with the base through a torsional spring. Its angle of rotation indicates the torque. It can be obtained from the bond graph via the lower s-junction through an activated bond with an integrating member (this is an element of block diagram where the output variable is the time integral of the input). The integrating member transforms the rotation frequency ω into the angle φ. The necessary condition for the activated bond is that the torque required by the indicator remains negligibly small. The bond graph contains also the moment of inertia of the casing. It is connected to the lower s-junction, whilst the moment of inertia of the generator shaft is modelled by the I-element at the upper s-junction.

The s-junction with the I- and C-element indicate the possibility of a resonance (see § 3.1), and Fig. 3.12 does not contain a damping R-element. In practice additional friction elements are important in order to eliminate or at least damp torsional vibrations of such

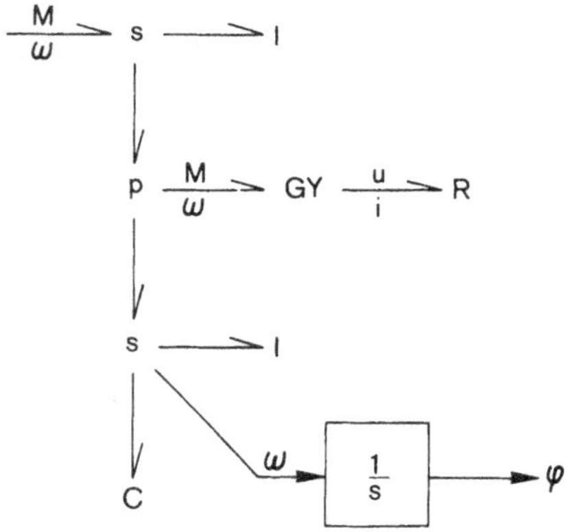

Fig. 3.12. Bond graph of a torque meter with electrical d.c. machine.

torque meters. In Fig. 3.12 the working R-element of Fig. 3.11 is replaced by a gyrator modelling the electrical generator (Chapter 5) and by an R-element dissipating the electric power.

The bond graph of Fig. 3.12 does not contain the torque reaction of the electric wires. These wires must be sufficiently flexible for high accuracy and, furthermore, the torque reaction of the cooling air of the generator must be suppressed. This is achieved in practice by having the cooling air enter through the plane of the axis of rotation or at least symmetrically to it.

It can be seen from the above examples that the setting up of mechanical bond graphs always requires that each member with a definite velocity or rotation frequency (shaft, casing) is associated to an s-junction. The efforts attack them through p-junctions on both s-junctions as shown on Fig. 3.11. The mass forces are an exception since they act on each component separately and must be applied to the s-junctions with the corresponding rotation frequency, as shown by the I-element of Fig. 3.12. It is implied in this procedure that rotation frequencies and velocities are measured with reference to absolute space. In other words the reference system must be a so-called inertial frame of reference.

Similar rules apply to linear displacements and forces, where for each velocity an s-junction is set up on which the forces act through p-junctions with the exception of the mass forces.

In order to set up bond graphs systematically it is also good practice to assign an s-junction to a fixed base and to apply all forces or torques through p-junctions. Then the velocity of the base is set equal to zero whence all adjacent bonds disappear (§ 2.4).

3.3.2. *Longitudinal oscillations of railway trains*

Figure 3.13 shows, as a further example, a schematic and a bond graph of a railway train consisting of a locomotive (left) and two cars, which are idealized as mass points with the mass m. The coupling consists of a spring with constant k and of a friction

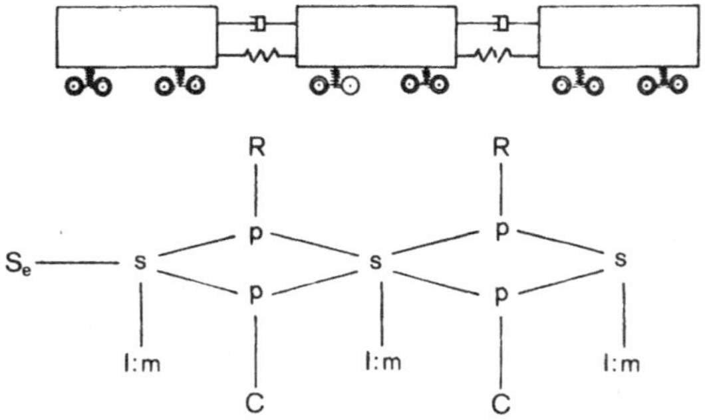

Fig. 3.13. Railway train with longitudinal oscillations, schematic (top) and bond graph (bottom).

member with the constant f. An effort source, e.g. periodic changes of the locomotive traction effort, can produce longitudinal oscillations. The bond graph is set up systematically as follows:

(1) Each car has a definite velocity corresponding to an s-junction, where an I-element for the mass is attached.

(2) The variable effort of the locomotive attacks the bond graph at the s-junction on the left. There is an equal reaction force from the driving wheels of the locomotive on the rails, but this is not included in the above bond graph since the rails are stationary in our frame of reference.

The forces of the spring and of the friction member in the couplings act on the adjacent cars equally, whilst the closing up (compression of the buffers) is determined from the difference of the speed of the cars. The corresponding R- and C-elements are consequently applied through p-junctions on the s-junction representing the speed of each car.

Figure 3.12 contains, furthermore, the half-arrows for the positive power directions. The next step is to set up block diagrams in the desired causality or to simulate the dynamic behaviour by the ENPORT programs.

3.3.3 *Systematic construction and simplification of bond graphs with moving reference frames*

Following on from the above examples we shall now treat the systematics of bond graphs for systems where the base is also moving, i.e. with moving reference frames. As mentioned, it is advisable to assign an s-junction to each member with a definite velocity, including the base. The mechanical elements are applied in between through p-junctions.

Figure 3.14 contains a simple drive with load similarly to Fig. 3.4 but without gear-reducer or compliant shaft. Both shaft and casing rotate with a certain frequency (or frequency zero) and are assigned an s-junction. The active elements are applied between them through p-junctions. The coupling of the shafts produces the top bond between the upper s-junctions and the common base the bond between the lower s-junctions. Both upper and lower s-junctions can be contracted (§ 2.4) leading to the bond graphs of Fig. 3.14 (centre).

Supposing now that the base does not rotate, the lower s-junction with its bond vanishes. Consequently, the remaining junctions each contain only two bonds and vanish, producing the simple bond graphs of Fig. 3.14 (bottom).

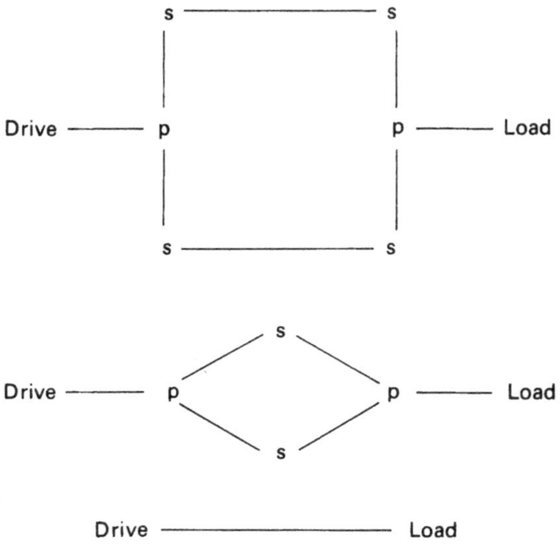

Fig. 3.14. Drive and load with rotating base (top) and simplified with fixed base (bottom).

Instead of assigning a zero rotation frequency to the base, one could postulate the shaft without rotation, whence the remaining bond contains the rotation frequency of the base with reference to the shaft. Generally it is advisable to assign a vanishing flow (rotation frequency) to the s-junctions with the greatest number of bonds in order to obtain the maximum of simplification. Usually, but not always, this is the base. Assigning, on the other hand, a rotation frequency of zero to a different s-junction, the remaining bonds contain the rotation frequency relative to the latter and to the components that rotate with it.

3.3.4. *Application to planetary gear-reducers*

The essential property of planetary gear-reducers is that the axis of some wheels, the planets, move around in space guided by a planet carrier. Such a reducer results, in principle, by allowing the casing of an ordinary gear-reducer to rotate and by coupling it to a third shaft in which the reaction torque appears. Planetary gears have usually the form of Fig. 3.15, but also automotive differentials with bevel gears belong to this class. We shall set up a bond graph for them as an example for moving reference frames.

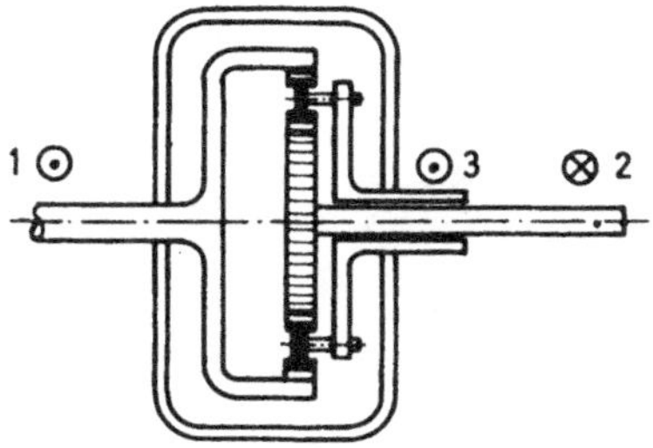

Fig. 3.15. Schematic of a planetary gear-reducer. The arrow-head ⊙ indicates movement out from the plane of the paper and the arrow tail ⊗ movement into this plane.

The planet carrier is the moving frame. If it is stopped, the device acts as a simple transformer or reduction gear with the factor i according to the following equation:

$$\omega_2 = i\,\omega_1 \qquad M_2 = \frac{M_1}{i} \qquad \text{if} \quad \omega_3 = 0 . \tag{3-4}$$

It should be noted regarding signs that with i positive it is supposed that the rotation frequency is positive if the upper half of the input shaft emerges from the plane of the figure and the upper half of the output shaft enters into this plane, as shown in Fig. 3.15 by arrow-heads ⊙ and arrow tails ⊕. The arrow-head of the planet carrier indicates that its direction of rotation is positive if the upper half comes out from the plane of the paper.

With rotating casing we shall at first consider both faces of the reducer separately and assign to it two s-junctions, one of which represents the rotation frequency of the casing (Fig. 3.16, left). Between the s-junctions there is transformer behind p-junctions (§ 3.3.2). They indicate that the reaction torque of both shafts acts on the casing. Both lower s-junctions refer to the rotation frequency of the common casing and are coupled by a bond.

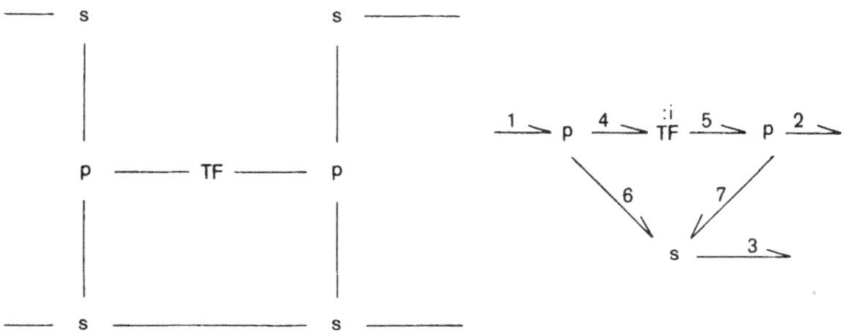

Fig. 3.16. Bond graph of a planetary gear-reducer.

Figure 3.16 (right) is obtained by simplification, i.e. by contracting both lower s-junctions and by elimination of the upper s-junction each carrying only two bonds. This part of the figure also contains the positive power directions. The bonds are consecutively numbered where the reference numbers 1, 2, 3 refer to the annular gear, to the sun gear, and to the planet carrier.

The directions of the positive energy fluxes are chosen in Fig. 3.16 such that the rotation frequencies ω_6 and ω_7 are deducted from the first and the second p-junction. Both are equal to the frequency ω_3. This corresponds to the directions of rotation of Fig. 3.15 as obtainable from a geometric consideration.

The bond graph of Fig. 3.16 is a power-conserving junction structure, since, at all times the junctions and the transformer conserve energy and power instantaneously. The losses will be represented separately, usually by R-elements. This bond graph corresponds to (3-4) and to the following equations implicit in the different junctions:

$$\begin{array}{lll} \text{1st p-junction:} & M_1 = M_4 = M_6 & \omega_1 - \omega_4 - \omega_6 = 0 \\ \text{2nd p-junction:} & M_5 = M_2 = M_7 & \omega_5 - \omega_7 - \omega_2 = 0 \\ \text{s-junction:} & M_6 + M_7 - M_3 = 0 & \omega_6 = \omega_7 = \omega_3 \end{array} \qquad (3\text{-}5)$$

The following relations between torque and rotation frequency in the three exterior shafts are obtained by elimination of the intermediate variables.

$$\omega_2 = i\,\omega_1 - (1 + i)\,\omega_3$$

$$M_3 = (1 + 1/i)\,M_2 \qquad M_2 = 1/i\,M_1 \tag{3-6}$$

In automotive differentials $i = 1$, and if ω_1 denotes the rotation frequency of the bevel gear carrier it results in $M_2 = M_1$ and $M_3 = 2\,M_1$.

For further simplification the transformer of Fig. 3.17 is drawn over the left junction whence a transformer with the reciprocal factor 1/i enters the lower bond (§ 2.4). In Fig. 3.17 (bottom) both p-junctions are combined which requires that the transformer carries the ratio 1/(1+i) according to § 2.4.

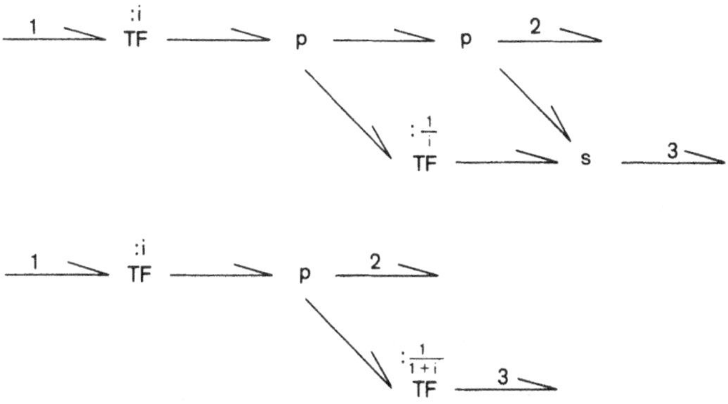

Fig. 3.17. Simplification of Fig. 3.16 leaving on bottom essentially only a p-junction.

Figure 3.17 (bottom) contains essentially only a p-junction. If, in particular, the factor i is equal to 1 as in automotive differentials the transformer on the left vanishes and the transformer on the right has the factor 2. Such a differential is the simplest representation of a p-junction in mechanics. The remaining transformer on the right can be eliminated by a further gear reduction if, for instance, in automotive differentials the drive shaft rotates with twice the frequency of the bevel gear carrier.

With planet gears the ratio i=1 cannot be reached because then the annular gear and the sun gear would need to have both equal pitch diameter and number of teeth.

Bond graphs constructed systematically represent the relatively complex relations of planetary gear-reducers by simple junctions and a transformer. The choice of positive power direction, and consequently of the signs, is by no means trivial, but important, and must correspond to the geometry of the device. Stopping one shaft, e.g. shaft 1 in Figs. 3.15 and 3.16, one obtains a reduction gear between shafts 2 and 3. Its relations are contained in (3-5) and (3-6) by setting $\omega_1 = 0$, whence M_1 is the reaction torque of the casing.

3.4. Perturbations, Losses, and Efficiencies

Bond graphs with junctions and transformers as described in §3.3 are power-conserving and consequently loss free. The real losses of a machine, such as the friction in the gears, are represented by R-elements or by sources driven in reverse (see below). For clarity it is convenient to divide the losses into two classes, although this is only normally done with hydrostatic machines and centrifugal pumps.

(1) The torque losses (effort losses) that reduce the output torque but do not influence the output frequency. They are modelled by an R-element connected with the machine through an s-junction (Fig. 3.18, left). The equations are

$$M_2 = M_1 - M_3(\omega_1) = M_1 - Z\omega_1 \qquad \omega_2 = \omega_1 \tag{3-7}$$

where in the linear case the rotative impedance Z is constant. Such losses are generally called effort losses since the flows are unaffected by them.

(2) The flow losses reducing the output frequency but leaving the torque unchanged. They are represented by an R-element through a p-junction (Fig. 3.18 right). Here the following equation applies:

$$M_2 = M_1 \qquad \omega_2 = \omega_1 - \omega_3 = \omega_1 - YM_2 \tag{3-8}$$

with the rotative admittance Y constant in the linear case. These losses are generally called flow losses or slip losses.

A gear-reducer has no slip losses because of the positive drive through the teeth. A belt-drive and other friction drives have slip losses and need an R-element behind a p-junction. Torque losses can be added for the various friction sources, especially bearing friction.

The R-elements for the losses are generally non-linear because the loss torques are not proportional to the rotation frequency but an arbitrary function of it. Linear R-elements are obtained only with

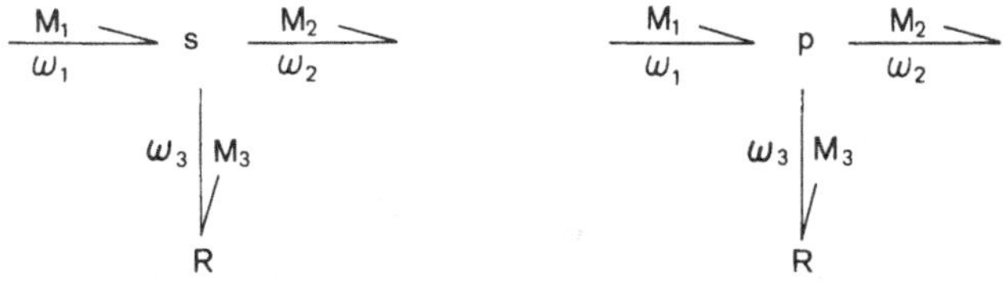

Fig. 3.18. Torque (effort) losses by an R-element behind an s-junction (left) and slip (flow) losses by an R-element behind a p-junction.

the so-called viscous friction in fluid power (with constant viscosity which is a poor approximation to reality), whilst with dry friction the loss torque is independent of rotation frequency but always opposed to its direction.

The R-elements well represent the physics of the losses since they correspond to recognizable single effects (slip, bearing friction), but in mechanical engineering the efficiency is frequently preferred. The efficiency is defined as the ratio of output power to input power and thus a dimensionless quantity smaller than 1. It is a multiplicator acting on the ideal loss free variables to determine the real output values.

Similarly as with the losses above, it is convenient to distinguish with the efficiency between:

(1) The effort efficiency caused by the loss torque

$$\eta_e = \frac{M_2}{M_1} = 1 - \frac{Z\,\omega_1}{M_1}\,. \tag{3-9}$$

(2) The flow efficiency caused by the slip

$$\eta_f = \frac{\omega_2}{\omega_1} = 1 - \frac{Y\,M_2}{\omega_1}. \tag{3-10}$$

The second form of (3-9) and (3-10) applies only to linear R-elements.

The product of effort and flow efficiencies is the total efficiency, i.e. the ratio of output to input power

$$\eta = \frac{\dot{E}_2}{\dot{E}_1} = \frac{M_2 \omega_2}{M_1 \omega_1} = \eta_e\,\eta_f\,.$$

Figure 3.18 and eqns. (3-9) and (3-10) are especially convenient with linear losses, e.g. if the loss torque is proportional to rotation frequency and independent of the useful torque. However, the loss torque is frequently independent of rotation frequency and proportional to the working torque. In this case the bond graph of Fig. 3.19 (left) is more useful. It contains an effort source for the loss torque controlled through an activated bond by the input torque M_1:

$$M_2 = M_1 - \epsilon M_1, \qquad \eta_e = 1 - \epsilon, \qquad \omega_2 = \omega_1. \tag{3-11}$$

Fig. 3.19. Torque and slip losses by controlled reverse-driven sources.

For slip losses proportional to rotation frequency and independent of torque the bond graph of Fig. 3.19 (right) is a convenient representation. It contains a controlled flow source proportional to the input frequency as shown by the gain ϵ of the block

$$M_1 = M_2 \qquad \omega_2 = \omega_1 - \epsilon\omega_1, \qquad \eta_f = 1 - \epsilon. \tag{3-12}$$

The gain of the block of Fig. 3.19 is in both cases equal to the complement of the corresponding efficiency. If this gain is independent of the operating variables, the corresponding efficiencies are constant.

The half-arrows of Fig. 3.19 show a positive power direction into the sources corresponding to the power dissipation by the losses. Contrary to other sources, the energy flux cannot reverse its direction, i.e. these sources cannot feed power into the system. From this it can be seen that loss processes are not reversible, as treated in more detail in § 6.3 with entropy generation. There the R-element is replaced by an RS-field.

The sources of Fig. 3.19 are an example of sources driven in reverse and absorbing power. This power is either dissipated, as in real losses, or stored or finally transmitted to some energy domain. The designation source indicates that in the case of an effort source the effort is independent of the flow for both directions of flow, and likewise for a flow source.

Figure 3.20 shows a bond graph with both effort and flow losses. This is obtained from a combination of both parts of Fig. 3.19 with a slightly different arrangement.

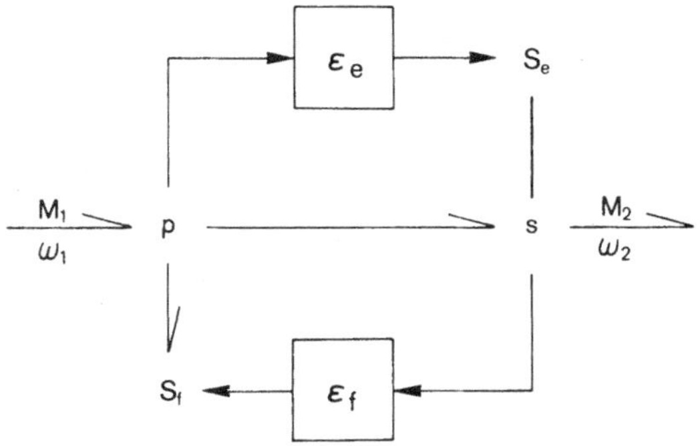

Fig. 3.20. Combination of both losses with controlled effort and flow sources.

Gears with teeth that are not cut perfectly produce periodic fluctuations of the rotation frequency. They can be represented according to Fig. 3.21 as a flow source behind an s-junction feeding into the output shaft. The s-junction is controlled by the angle φ determined through the s-junction and the activated bonds by integration with respect to time. For other perturbations, e.g. rolling of rough surfaces, the noise generators known from electronics can be used. They produce a statistical perturbation signal corresponding to the roughness of the surface.

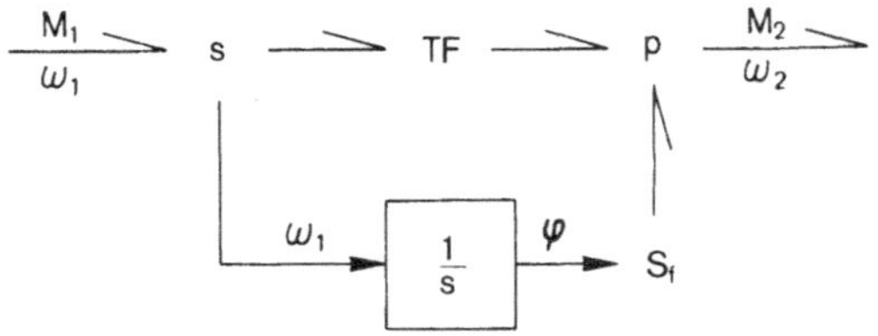

Fig. 3.21. Representation of fluctuations of rotation frequency of gears by a controlled flow source.

3.5. Couplings and Transmissions

Couplings of any kind including friction couplings, hydrokinetic, electromagnetic, and hydrostatic couplings (see below) produce a controlled slip and are essentially free from loss torques. They are modelled by a controlled R-element with a p-junction as shown on Fig. 3.22 (left). The control signal acts upon the R-element through the activated bond.

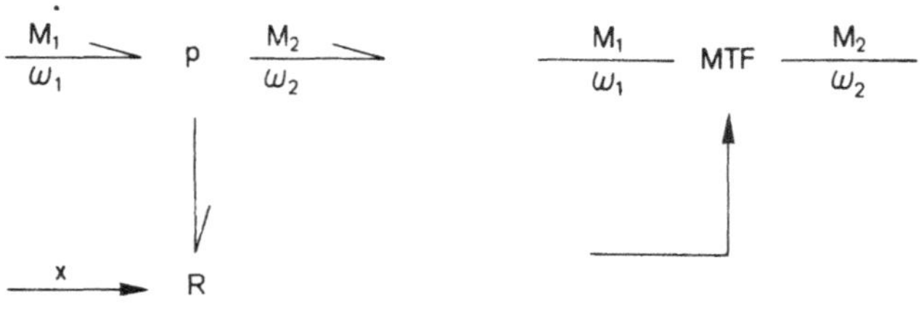

Fig. 3.22. Coupling as R-element in parallel (left), transmission as modulated transformer (right).

Figure 3.22 (left) contains an ideal coupling with no losses except the controlled parallel resistance necessary for the operation. The additional small torque losses experienced in practice, e.g. in the bearings, can be represented by R-elements in series. In order to disengage the coupling, the mechanical impedance of the R-element must be made to zero, so that the transmitted torque vanishes. Rest values of the mechanical impedance give rest torques usually negligible with friction couplings but a major design criterion with other kinds of couplings. The output torque is essentially equal to the input torque since couplings produce no torque of their own. The slip losses produce a corresponding thermal energy flux (more accurately entropy flow, § 6.3), which must be dissipated. If the coupling is able to dissipate the entire thermal energy or heat production it can be used as a control coupling.

Contrary to a coupling, a transmission acts as transformer and a variable speed transmission as a modulated transformer (Fig. 3.22, right, with the control signal x). It produces an additional torque if the output frequency is smaller than the input frequency. Apart from more complex design, a transmission has generally higher losses – both torque and slip losses in friction wheel and other transmissions. Even if the output torque of the transmission is limited to a constant value independent of output frequency, it needs a smaller input torque for reduced output frequency. With a coupling, on the other hand, the input torque is always equal to the output torque as mentioned.

The difference of hydrokinetic couplings and transmissions is easily determined from the design. The hydrokinetic transmission needs a fixed crown of blades to produce the additional torque. These blades produce losses of their own in addition to the losses experienced in a hydrokinetic coupling.

In summary, couplings comprise essentially a parallel resistance for controlled slip, whilst a variable transmission is a modulated transformer generating new torque by its own.

Figure 3.23 contains a bond graph for a friction coupling with appreciable actuating force. It is actuated through the bond 3 from the left and the C-element models the spring of the coupling. The coupling impedance, i.e. the R-element, is modulated by the position x_3 of the actuator rod which is obtained from the integrating member, the block with the gain $1/s$.

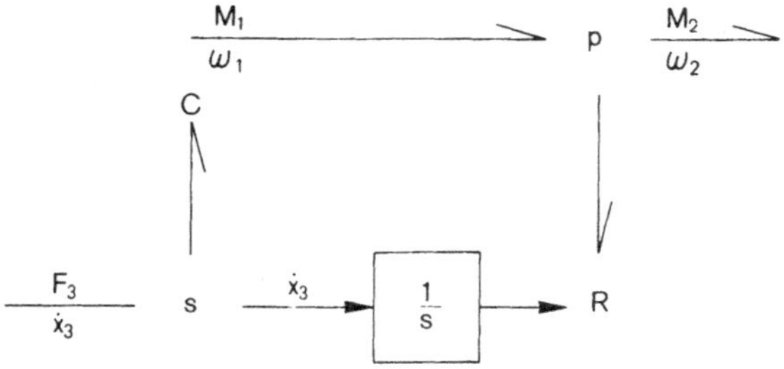

Fig. 3.23. Friction coupling (clutch) with actuating forces. The C-element models the spring pressing the friction faces together.

Bond graphs give a vivid representation of the fundamental differences between couplings and transmissions.

CHAPTER 4

Fluid Mechanical Applications

4.1. Hydrostatic Energy Transport

This chapter treats the hydrostatic energy flux and its conversion into other energy domains by class transformers as described in § 2.2. Bond graphs have their prime application only if the power flux is equal to the product of an effort and a flow variable to a sufficient approximation. Therefore they are limited to incompressible or almost incompressible fluids, which includes oil hydraulics (oil as working fluid) but excludes pneumatics due to the high compressibility of air. Furthermore, fluid-flows with free surface are excluded. Limiting the considerations to small deviations from an operating point bond graphs can also be useful for gases or air.

In incompressible flows the total pressure is the sum of the static and the dynamic pressure as follows:

$$\dot{E} = \dot{V}\,(p_{stat} + \frac{\rho}{2}\,v^2) = \dot{V}\,p_{tot}$$

where ρ is the mass density and $\dot{V}$ the volume flow.

Consequently the total pressure p_{tot} is a suitable effort variable for bond graphs. Convergent and divergent pipes without losses have no effect, whilst losses due to a sudden increase of cross-section are modelled by R-elements. There are some irregularities with accelerated flows and with changes of flow section, but these are outside the scope of this book.

Alternatively, it is possible (after Karnopp, ref. 3, pp. 222–9) to use as effort variable the static pressure which is responsible for the stress on the pipe walls. On changes of the pipe section the static pressure becomes smaller (narrowing) or larger (widening) according to Bernoulli as modelled by

R-elements. This reference also contains interesting considerations about moving reference frames, e.g. in turbines.

We shall use the total pressure as effort variable, but the difference is usually negligible in fluid power and in lubrication theory (§ 4.2).

4.1.1. *Fundamental components and circuits*

R-elements represent the different pressure drops in fluid power, either intentional in throttles or undesirable in pipes. The compressibility of the oil is modelled by C-elements as well as the so-called hydropneumatical accumulators (ref. 4) which are interesting non-linear components. I-elements for the mass action of the fluids are used relatively rarely.

For a circuit with a restrictor between two pipes with different pressures (Fig. 4.1), we observe that the volume flow is equal in both pipe-joints of the restrictor and only a function of the pressure drop. Consequently the restrictor is an R-element applied through an s-junction to two p-junctions that connect with the main pipes, as shown on Fig. 4.1 (right).

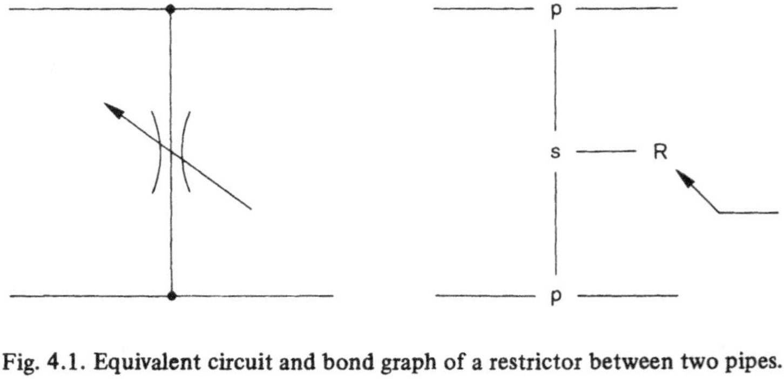

Fig. 4.1. Equivalent circuit and bond graph of a restrictor between two pipes.

From these examples we obtain the following rules for the construction of bond graphs from fluid power circuits. They are applicable also for electronics by replacing pressure and volume flow by voltage and current respectively:

(1) A p-junction is associated to each point with a definite pressure, especially to each branching point of pipes.
(2) Components between two pipes are applied between corresponding p-junctions through an s-junction. This is only possible if the volume flows in both pipe-joints of the component are equal.

The volume flows of both pipe-joints are equal with R-elements, I-elements, and with hydrostatic pumps and motors. They are also equal in hydraulic cylinders having equal piston areas on both sides, but this excludes the so-called differential area cylinders. The C-elements, including the hydro-pneumatic accumulator, have only one pipe-joint and act between the absolute space and the pipe to which they are connected by a p-junction. An exception is the compressibility of the oil in the main pipe including its increase of diameter under pressure.

Figure 4.2 shows a hydraulic cylinder with equal areas (top left) as schematic corresponding to the international circuit symbol. The

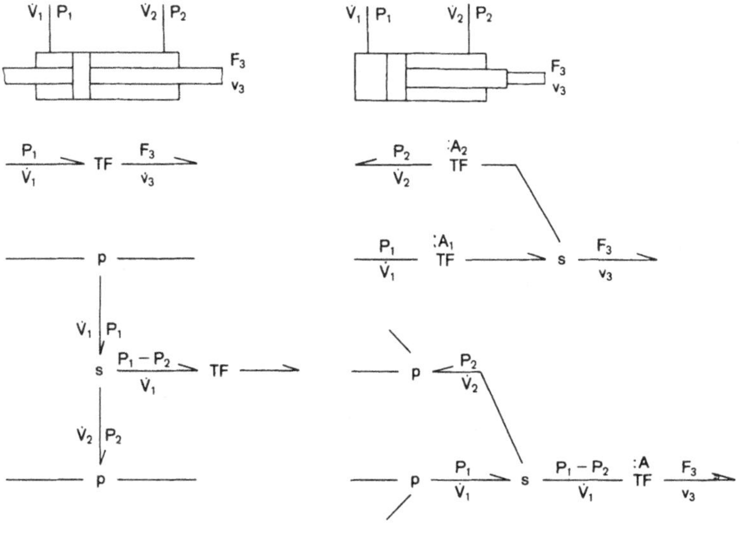

Fig. 4.2. Hydraulic cylinders with equal (left) and different (right) areas, schematics, and bond graphs.

bond graph in the centre with the transformer is valid if there is pressure only in one pipe, whilst for pressures in both pipes 1 and 2 the lower bond graph with the transformer and the junctions is applicable. Here the volume flows are equal and the net force is equal to the pressure difference times the piston area, justifying the connections through an s-junction. The half-arrows for the power direction indicate that both piston areas act one against the other.

Figure 4.2 (right) shows a differential area hydraulic cylinder, where the areas A_1 and A_2 are different. The top contains a schematic, the centre a bond graph with two opposed transformers connected through s-junctions with the piston rod. At bottom right both piston areas are supposed equal whence the s-junction can be interchanged with the transformer, leading to the former representation at bottom left. It is seen that with equal piston areas the force depends only on the pressure difference in the pipes.

Figure 4.3 contains two bond graphs, on the left a modulated transformer for an ideal hydrostatic variable displacement pump as

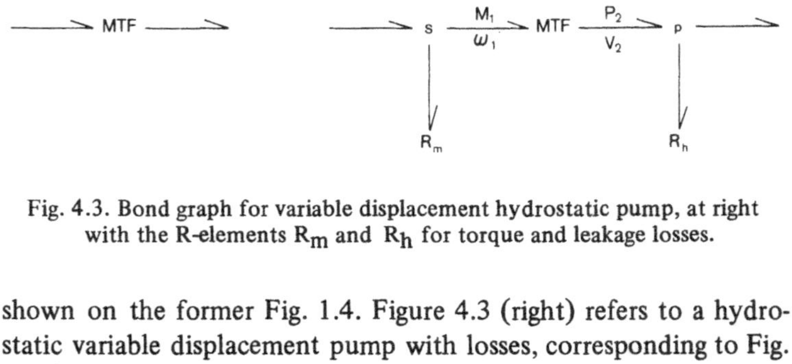

Fig. 4.3. Bond graph for variable displacement hydrostatic pump, at right with the R-elements R_m and R_h for torque and leakage losses.

shown on the former Fig. 1.4. Figure 4.3 (right) refers to a hydrostatic variable displacement pump with losses, corresponding to Fig. 1.4 (right), with a mechanical R-element for the torque losses and a hydraulic R-element (denoted R_h) for the leakage. They belong to the class of R-elements since they give an arbitrary linear or quadratic relation between flow and pressure drop. The torque losses act with the same rotation frequency as the pump and are additive to the useful torque, and are consequently connected through the s-junction on Fig. 4.3, in accordance with § 3.3.

Figure 4.4 shows a hydrostatic pump operating between two pipes 1 and 2 with arbitrary pressures. It has no torque losses, but internal leakage between both pipes and an external leakage between pipe 1 on top and the tank. The equivalent circuit at left represents the leakage by two restrictors, the bond graph at right by R-elements, denoted R_i and R_e.

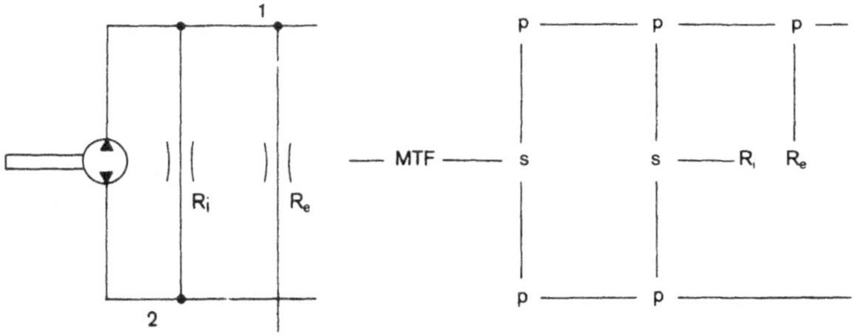

Fig. 4.4. Hydrostatic pump between two pipes with internal and external leakage.

Bond graphs for all kinds of hydraulic circuits can be easily set up from the method given by the above examples. Their combination with other parts of a multienergy domain or interdisciplinary system presents no difficulty.

4.2. Hydrostatic Bearings

Hydrostatic bearings, where mechanical and hydrostatic variables co-operate, are well suited to bond-graph representation. Figure 4.5

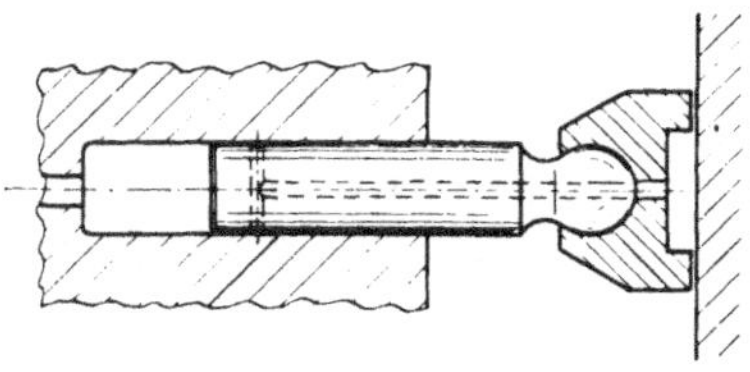

Fig. 4.5. Piston and slipper as example of an hydrostatic bearing.

shows a slipper and piston combination (frequently used in axial piston hydrostatic pumps), where the slipper abutes on the wall on the right and is loaded by the piston in the cylinder under pressure on the left. High-pressure oil is taken through the diametral clearance between the cylinder and piston head into an annular grove. Subsequently it is ducted through the bore into the central recess of the slipper for lubrication. There is a definite pressure drop between the cylinder and the annular grove due to the small clearance which has a high impedance.

Figure 4.6 (top) shows a schematic of a slipper similar to Fig. 4.5 an equivalent circuit (left) and a bond graph (bottom right). The restrictor with the impedance Z_1 between the supply from a pressure source and the slipper itself is important for the operation, as will be seen below.

It is important to recognize that the slipper incorporates the following two actions:

(1) The gap between the slipper and support acts like an impedance modulated by the gap height, as given from the so-called gap formula (ref. 4)

$$Z = \frac{p_0}{\dot{V}} = \frac{12\mu L}{wh^3}, \tag{4-1}$$

where μ is the viscosity, L the sealing length, w the sealing width, and h the gap height. It is seen from (4-1) that the admittance (the reciprocal impedance) increases with the third power of the gap height h.

(2) The slipper acts like a transformer similar to the hydraulic cylinder of Fig. 4.1, transforming pressure and volume flow into force and velocity equal to the time derivation of the gap height $\dot{h}$:

$$F = A_m\, p_0 \qquad \dot{V} = A_m\, \dot{h}, \tag{4-2}$$

where p is the pressure in the recess and A_m the mean area of the slipper.

The above actions work in parallel and are represented by the

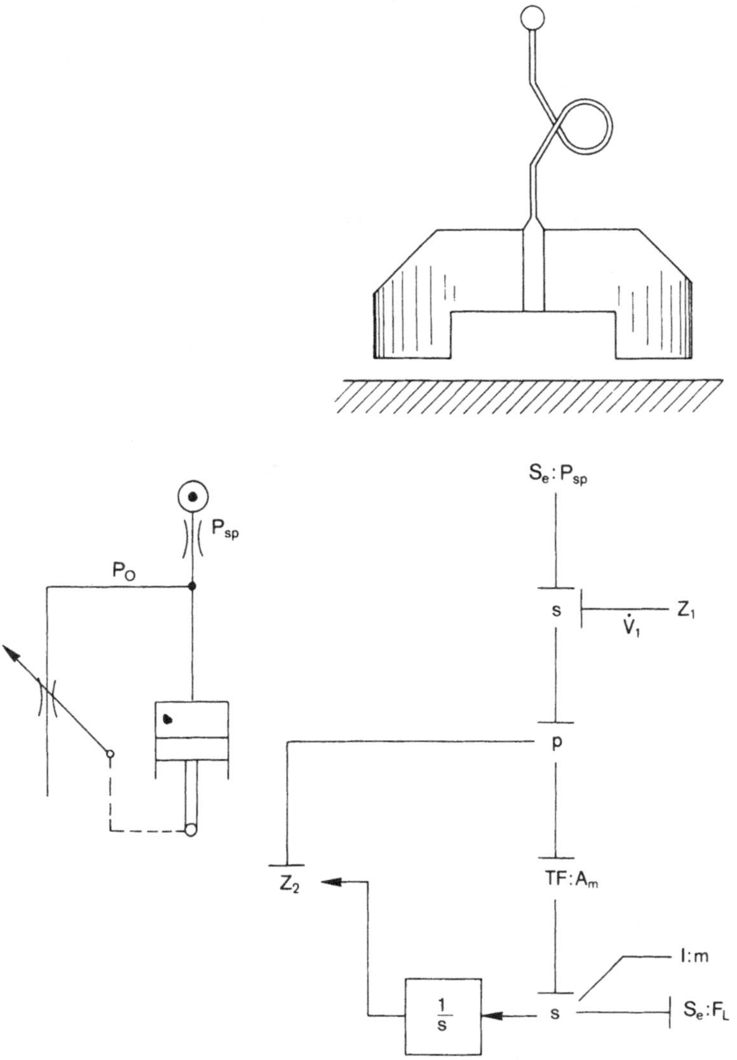

Fig. 4.6. Schematic (top right), equivalent circuit (left), and bond graph (bottom right) of an hydrostatic slipper.

equivalent circuit on Fig. 4.6 with a modulated impedance. The bond graph (bottom right) has the upstream R-element Z_1 behind an s-junction, followed by the modulated impedance Z_2 and the transformer through a p-junction. The lower s-junction determines the gap height together with an I-element for the mass of the slipper with piston and an effort source for the load (piston force).

Figure 4.7 contains the block diagram in the causality of Fig. 4.6. The operation can now be described as follows.

The supply enters the recess through the upstream restrictor Z_1 and lifts the slipper from the support against the load. This increases the admittance of the gap and its volume flow, producing a pressure drop at the upstream restrictor. The pressure drop increases and the pressure in the recess falls until the corresponding lifting force is in equilibrium with the load.

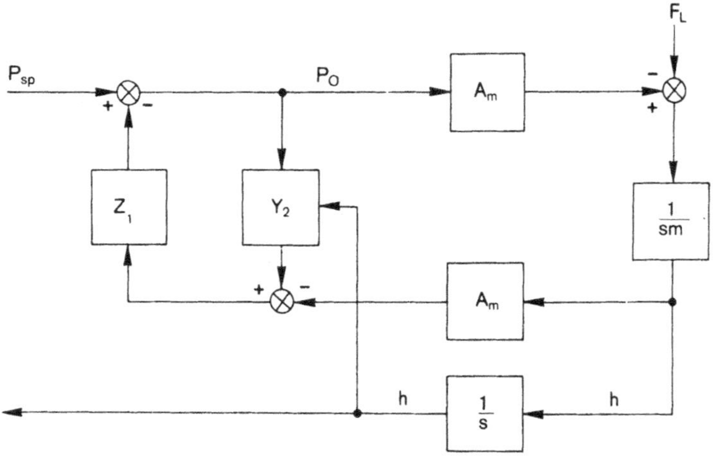

Fig. 4.7. Block diagram to Fig. 4.6.

The slipper incorporates thus an automatic control of the gap height against variations of the load as visible also on the block diagram of Fig. 4.7. The minimum gap height necessary for safe operation is a question of the physics of the contact between the sliding surfaces. The bond graph can give no information about it since it only displays the interaction of the variables due to the fluid mechanical actions.

Figure 4.8 shows a bond graph (left) and a block diagram (right) including the compressibility of the oil in the recess by the C-element with the value V/B (volume = V, B = compressibility

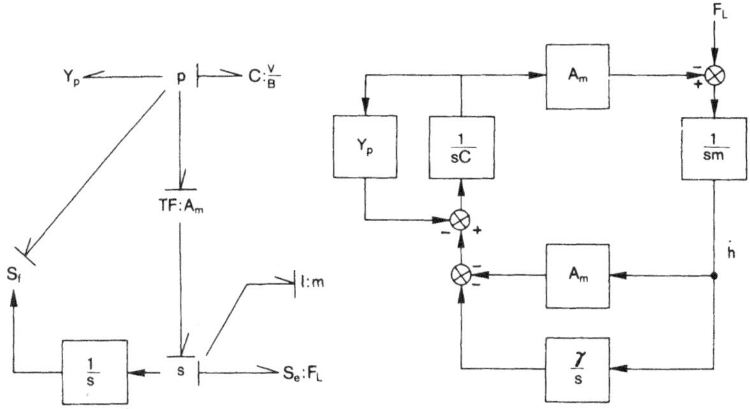

Fig. 4.8. Bond graph (left) and block diagram (right) of the slipper for small deviations from the operating point. Apart from the different causalities, the C-element for the compressibility of the oil in the recess has been added.

modulus). It is valid for small deviations from the operating point for which the volume flow at the modulated impedance can be expanded as follows:

$$\Delta \dot{V} = \dot{V}'^{p}_{-h}\,\Delta p + \dot{V}'^{h}_{-p}\,\Delta h = Y_3\,\Delta p + \gamma\,\Delta h. \qquad (4\text{-}3)$$

Equation 4-3 defines the admittance Y_3 and the gain γ. The admittance is added to the admittance of the upstream restrictor to form the combined admittance $Y_p = Y_1 + Y_3$. Thus (4-3) replaces the modulated impedance of the gap height by a modulated flow source and a parallel admittance. The block diagram of Fig. 4.8 implies a resonance as can be shown more distinctly by applying the formula of Mason (§ 1.3).

A remarkable factor which arises from the treatment of hydrostatic bearings by bond graphs is that the so-called squeeze film effects, i.e. the displacement of oil out from approaching sliding surfaces, are obtained automatically. They are contained in the action between the flows of the transformer in the lower block A_m of Figs. 4.7 and 4.8.

To be complete we still indicate the transfer function between the gap height h and the load force F in Fig. 4.8 as obtained by Mason's formula from the block diagram:

$$\frac{h}{F} = \frac{Y_p}{\gamma A_m} \; \frac{1 + \tau_1 s}{(1 + \tau_1 s)\, s^2/\omega_r^2 + \tau_2 s + 1} \, .$$

With the intermediate variables

$\omega_r^2 = \gamma A_m / Y_p m = k_{st}/m$ the resonance frequency with k_{st} the static stiffness of the bearing;

$\tau_1 = c/Y_p$ the time constant of the capacity of the recess with the parallel admittance Y_p;

$\tau_2 = A_m/\gamma$ a time constant built from the area of the recess and the gain of the flow source.

Figure 4.9 is the cross-section through a hydrostatic flow amplifier, also called hydrostatic transistor. This component is closely related to hydrostatic bearings, and Fig. 4.10 contains its

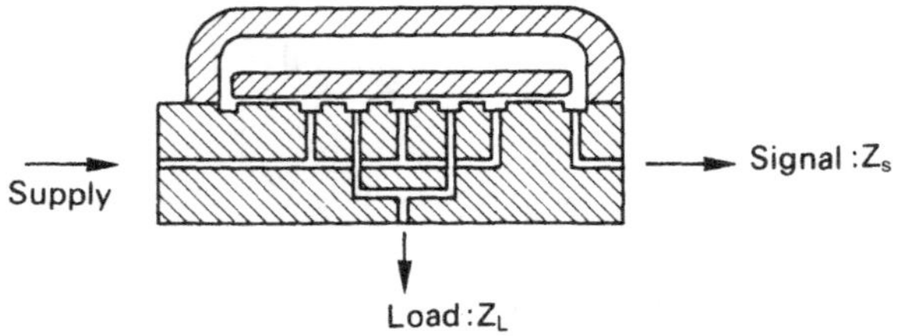

Fig. 4.9. Cross-section of a hydrostatic flow amplifier.

equivalent circuit (left) and the bond graph (right). The plate floats above the support with two kinds of recesses, one connected to the supply (Fig. 4.9, left) and the other one to the load. Hydrostatic gaps similar as in hydrostatic bearings are formed between both recess and the external annular grove. These gaps are an impedance modulated by the gap height, whilst the area in front and behind of the plate are transformers acting one against another. This is represented in Fig. 4.10 by the counteracting hydraulic cylinders and by the transformers acting on the s-junction on the right. The s-junction still contains an I-element for the mass and controls the gap height through the integrating block.

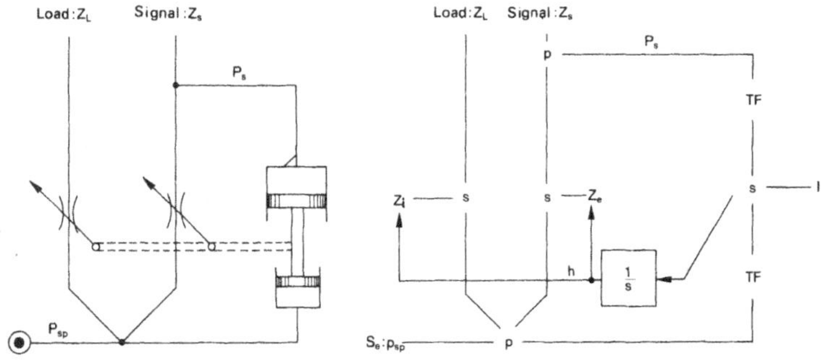

Fig. 4.10. Equivalent circuit (left) and bond graph (right) of the flow amplifier.

It controls the gap height through the impedance Z_e. If the volume flow at the signal port increases, or the signal impedance and consequently the pressure p_s decreases, the gap height is increased automatically such that the pressure remains approximately constant. This increases concurrently the height of the internal gap allowing more flow to the load port. With suitable dimensioning up to 50 flow gains between signal and load port can be obtained.

The example of the hydrostatic flow amplifier shows the usefulness of the combined representation by schematic, equivalent circuit, and bond graphs. For further calculation, especially for dynamic effects, it is most convenient to use the bond graph with suitable time-dependent elements.

4.3. Hydrostatic Control Systems and Drives

4.3.1. *Servocontrols*

In this section we shall introduce hydrostatic servocontrols and drives including hydrostatic couplings as examples for the common description and modelling of combined hydrostatic and mechanical systems by bond graphs.

Figure 4.11 contains a spool valve in a cylindrical bore with high-pressure oil supply in the centre, oil return to tank on both ends, and the ports A and B to a working cylinder below. These ports are

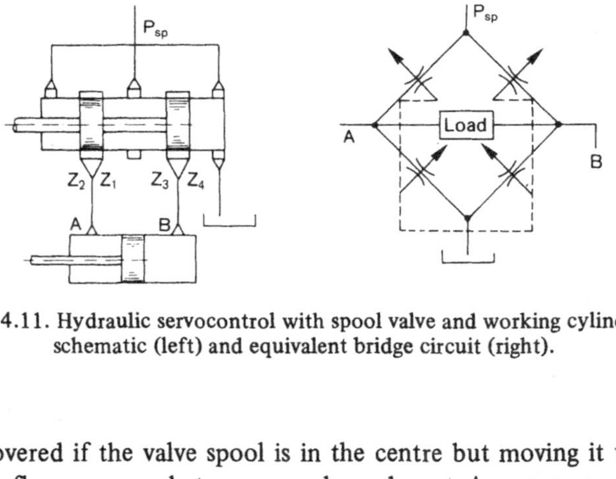

Fig. 4.11. Hydraulic servocontrol with spool valve and working cylinder; schematic (left) and equivalent bridge circuit (right).

just covered if the valve spool is in the centre but moving it to the left; a flow passage between supply and port A opens as well as another passage between port B and tank. With movement of the spool to the right, flow passages are opened between supply and port B and between port A and tank return. Modulated impedances are set up between the edges of the spool valve and the ports A and B, the so-called metering edges. Since we have four metering edges this type is also called four-way control valve.

Figure 4.11 shows the equivalent circuit on the right with the four impedances Z_1 to Z_4 drawn in a form of a bridge. The bond graph of Fig. 4.12 is obtained according to rules of § 4.1 simply by allocating a p-junction to each pipe branch and by putting the impedances and the intervening load behind s-junctions.

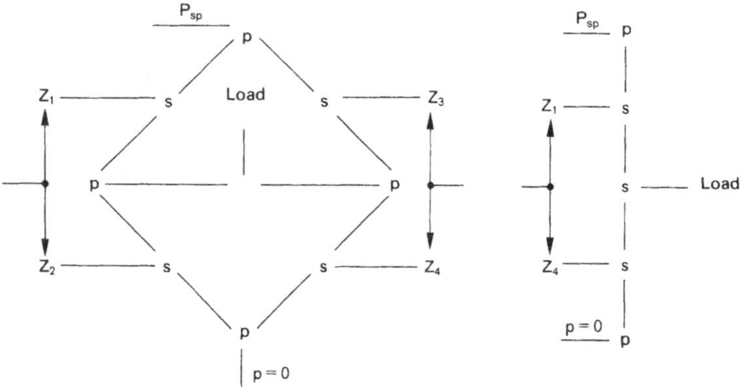

Fig. 4.12. Bond graph for the servocontrol of Fig. 4.11; left with four controlled impedances, right retaining only two working impedances with open metering edges.

If the spool moves to the right or left, each time only two flow passages are open and have a definite impedance, whilst the other metering edges are closed and do not carry any flow. Therefore the ineffective impedances are omitted in Fig. 4.12 (right) and only the working impedances Z_1 and Z_4 are retained for the case of a movement of the spool to the left. This bond graph is the starting point for discussing a hydrostatic servocontrol, but it can still be simplified by contracting the s-junctions. Also if the pressure in the tank vanishes, the lower p-junction can be eliminated. The relation between volume flow $\dot{V}$, load pressure p_1 and supply pressure p_{sp} is generally proportional to the displacement x of the spool and to the root of the pressure drop, although at very small values the flow is in fact directly proportional to pressure drop:

$$\dot{V} = cx \sqrt{p_{sp} - p_L} \quad (4\text{-}4)$$

Equation (4-4) applies to turbulent pressure drop at the metering edges and is used frequently as practical approximation.

A hydraulic control system, e.g. a copying attachment for machine tools in production engineering, comprises in principle a spool valve and an actuator similarly as in Fig. 4.11. The cylinder can be replaced by a hydrostatic motor.

Figure 4.13 shows the equivalent circuit of a servocontrol with spool valves and hydrostatic motor in AHEM (in the United States with NFPA) symbols with supply from a pressure source above. The spool valve is represented by the two effective impedances modulated by the control signal x. Compared to Fig. 4.11 the following perturbation effects have been added:

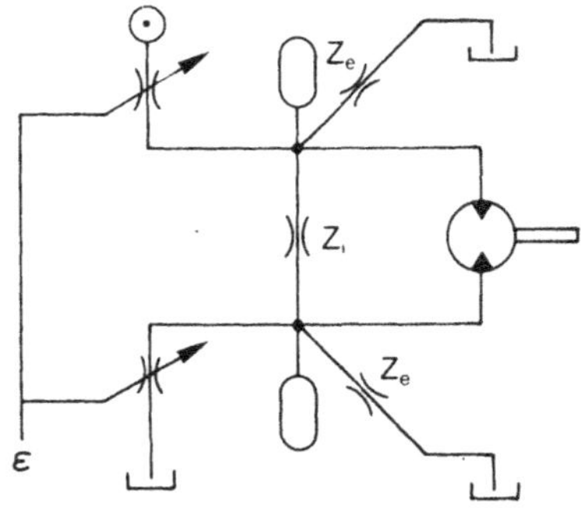

Fig. 4.13. Equivalent circuit of the servocontrol with accumulators for oil compressibility.

(1) The compressibility of the oil between the spool valve and the hydrostatic motor is represented by hydraulic accumulators.
(2) The impedance Z_i models the internal leakage of the hydrostatic motor between its pipe connections.
(3) The external leakage of the motor returning from the pipes to the tank is represented by the impedance Z_e.

Figure 4.14 contains the corresponding bond graph and is still recognizably similar to Fig. 4.13. The leakage impedances and the accumulators (C-elements) attack through p-junctions. Before augmenting the bond graph and calculating the different variables the following simplifications will be introduced:

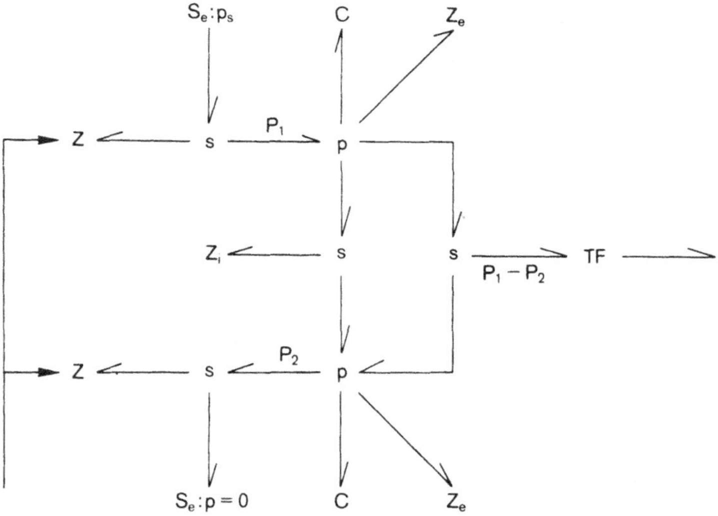

Fig. 4.14. Bond graph corresponding to Fig. 4.13.

(1) The external leakage is negligible and both capacitances are equal, a good approximation with hydrostatic motors but not so with cylinders, if the working piston is off centre. Then the leakage depends only on the pressure drop in the hydrostatic motor, and both hydraulic capacitors can be replaced by a common C-element of half its value. This comes from the series connection of the capacitors and can be determined by a simple calculation. Both metering edges of the spool valve are in series and can be replaced by one R-element with the double impedance. The resulting bond graph on Fig. 4.15 is valid for signals of arbitrary value.

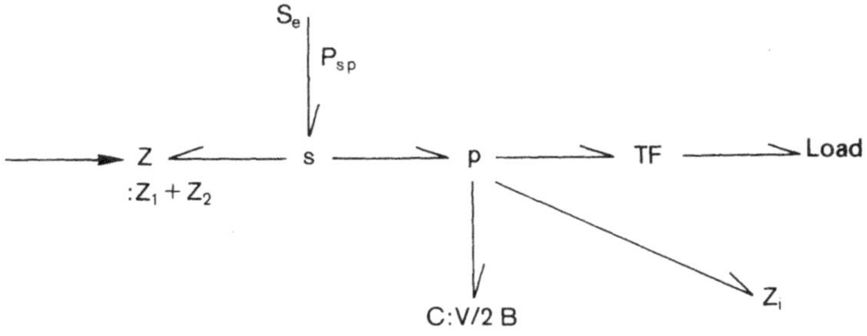

Fig, 4.15. Simplification of Fig. 4.14 with equal capacities and vanishing external leakage.

(2) Limiting now to small deviations from an operating point the general valve characteristic can be expanded as follows:

$$\dot{V}_L = kx - \sigma p_L \tag{4-5}$$

where $\dot{V}_L$, x, and p_L denote now these deviations. The gains k and G are determined from the characteristics.

The first term on the right-hand side of (4-5) is a controlled flow source from which the second term, a volume flow proportional to pressure, must be deducted as represented by the bond graph of Fig. 4.16 (left). Solving instead (4-5) for p_L one obtains the effort source with parallel R-element as shown on Fig. 4.16 (right).

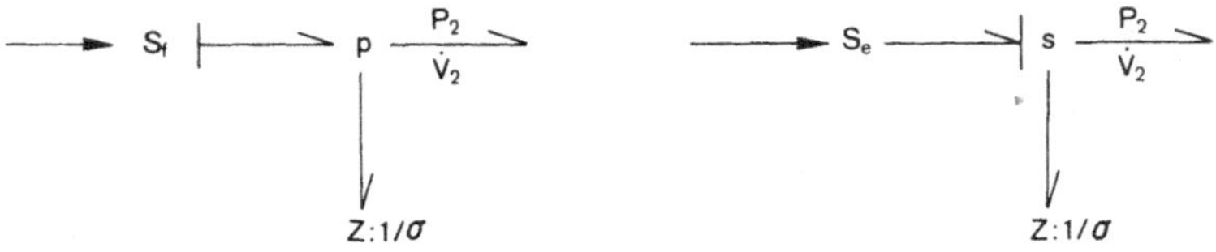

Fig. 4.16. Action of the spool valve with small deviations from the operating point (left) as controlled flow source with parallel R-element (right) as controlled effort source with series R-element

Both the representation with the flow source and with the effort source are useful in practice, since the sources impose a certain causality to the bond graph (§ 2.3).

Figure 4.17 contains bond graphs with different causalities obtained by combining Figs. 4.15 and 4.16. At the same time the load is modelled by an I-element for the moment of inertia (or the mass in the case of a cylinder), by an R-element for the friction, and by a force source for the variable load torques, e.g. due to cutting forces in machine tools. It is immediately visible from the bond graph that the C- and I-elements give a possibility of resonance that is damped by the different R-elements. This conclusion could also be confirmed by the corresponding block diagram.

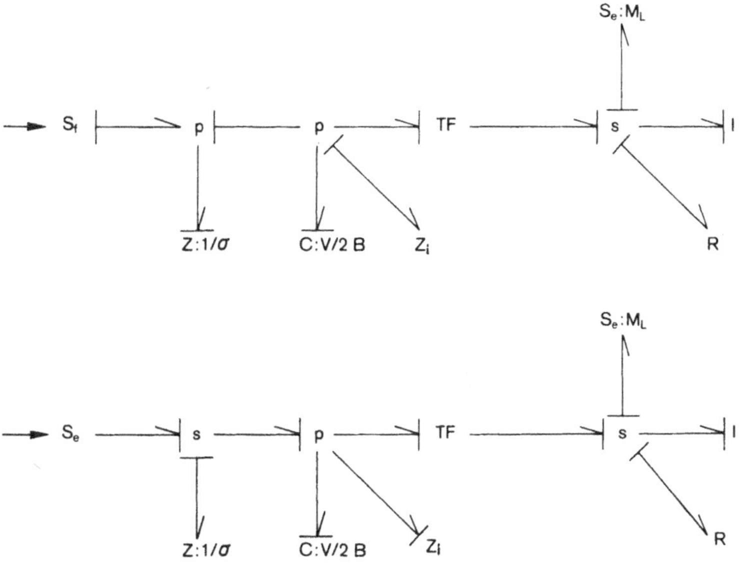

Fig. 4.17. Bond graph for the servocontrol in different causalities obtained by combining Figs. 4.14, 4.15, and 4.16.

4.3.2. *Hydrostatic transmissions and diesel engines*

Similarly to § 2.1 we describe here a bond graph for a drive system consisting of Diesel engine, hydrostatic variable displacement pump and motor, and a load. The pump and motor build a hydrostatic transmission in which the energy flux is given by pressure and volume flow. Strictly speaking there is a forward and a return of the working fluid as could be represented by separate bonds similarly to Figs. 4.13 and 4.14. For simplicity we suppose that the return line has no pressure whence the corresponding bond disappears.

Figure 4.18 (top) contains a word bond graph replaced in the lower part of the figure by standard elements as follows:

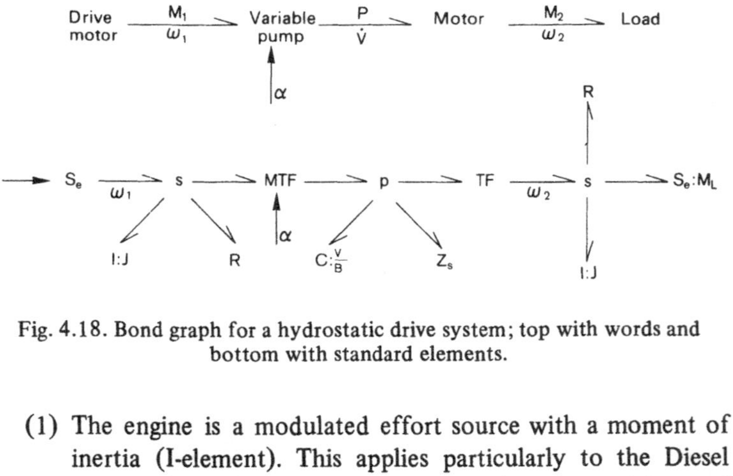

Fig. 4.18. Bond graph for a hydrostatic drive system; top with words and bottom with standard elements.

(1) The engine is a modulated effort source with a moment of inertia (I-element). This applies particularly to the Diesel engine where the torque is proportional to the position x of the injection rod but independent of rotation frequency. The accelerating torque depends on the rotation frequency of the engine and has to be deducted from its torque as correctly represented by the s-junction. The R-element (bottom left) indicates the torque losses of the hydrostatic pump and the Diesel engine including, for example, its cooling fan.

(2) The variable displacement pump is a transformer modulated by the signal α, given in the case of axial piston pumps by the inclination of a plate, the so-called swash-plate.
(3) The leakage of the pump and the motor is modelled by the R-element with Z_s in the centre and the compressibility of the oil by the C-element both connected through a p-junction.
(4) The hydrostatic motor is a fixed transformer.
(5) The load consists of an I-element for its moment of inertia and an R-element for the friction including loss torques of the hydrostatic motor and of a torque source indicating the absorption of the working torque. The latter is a reverse driven source as described in § 3.4.

Figure 4.18 contains several possibilities of resonance between both I-elements and the C-elements damped by the different R-elements. The engine can be controlled both by the control setting α of the hydrostatic pump and the position x of the injection pump. This leads to interesting combinations, allowing control either for minimum fuel consumption or minimum air pollution. In the simplest case the rotation frequency of the engine is maintained constant by controlling the injection, whence the output frequency is proportional to the control signal α. This and other control systems can be easily studied from the bond graph of Fig. 4.18.

4.3.3. *Hydrostatic couplings*

A hydrostatic coupling is obtained by connecting the casing of a hydrostatic pump not to the base but to another shaft, i.e. the output shaft, whilst the pump shaft is coupled to the input. The volume flow of the pump returns to the tank through a separate restrictor. If this restrictor is open, no pressure builds up and very little torque – only the loss torque is transmitted. Closing the restrictor causes pressure to build up and the reaction moment of the pump casing is transmitted to the output shaft.

Figure 4.19 contains a schematic (left) and a bond graph (right) according to the rules for moving reference frames in § 3.3. An s-junction is associated to the rotation frequency of both shaft and casing and the pumping action is modelled by the transformer behind the p-junction. The upper R-element represents the loss torque of the hydrostatic pump acting between shaft and casing. On the hydraulic side of the transformer there is the modulated restrictor Z_r and in parallel with it the leakage of the hydrostatic pump with the impedance Z_S.

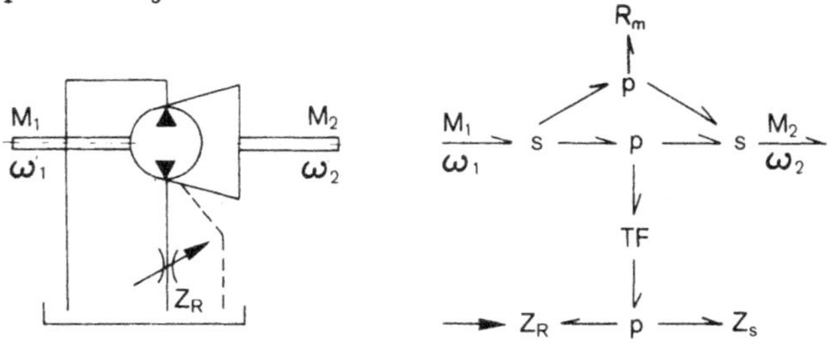

Fig. 4.19. Hydrostatic coupling with pump with rotating casing delivering against an adjustable restrictor; schematic (left) and bond graph (right).

The bond graph (Fig. 4.19) displays the following properties of hydrostatic couplings:

(1) The combination of transformer and hydraulic impedance acts similarly as a modulated mechanical R-element or mechanical coupling of Fig. 3.22. With vanishing hydraulic impedance no torque is transmitted, and with a completely closed restrictor or infinite impedance a small slip remains due to the leakage with its impedance Z_S.

(2) The torque losses between shaft and casing produce no losses with closed coupling. Consequently the input torque always equals the output torque except for small bearing losses that could be represented by separate R-elements.

(3) Any desired relation between torque and slip can be obtained by suitable pressure/flow characteristics of the controlling restrictor. In this sense a flow control valve (§ 4.4) produces a slip largely independent of torque, and a pressure relief valve limits the transmitted torque.

The losses of the hydrostatic pump appear as slip with closed coupling (due to leakage), and as rest torque with open coupling due to the loss torque. The output torque equals the input torque as in all couplings. In other words, the coupling cannot produce any torque, contrary to a variable transmission that acts like a modulated transformer. The thermal power (heat) corresponding to the losses appears in the oil from which it can easily be cooled away and not at the friction linings as in clutches. The hydrostatic coupling is consequently a control coupling supporting continuous slip.

Hydrostatic couplings have many advantages for drives that do not need torque increase at low output frequency. The main difficulty is the transfer of the pressure oil or of the control signal into the revolving casing for which the special transfer seal described in ref. 5, p. 49, is found to be a solution.

4.4. One-ports in Fluid Power Engineering

One-ports are components having only one bond; in fluid power only two pipe connections (§ 2.2). This includes standard restrictors which are simple R-elements, but in this section we shall describe direct acting and piloted pressure relief valves and also the so-called flow control valves. Such valves are frequently used for fluid power controls in production machinery and machine tools.

Figure 4.20 contains a schematic (top left) and the AHEM symbol (bottom left) of a simple relief valve. If the force due to the pressure exceeds the preload of the spring, the valve body lifts and opens a flow passage. The bond graph (right) represents the operation by the impedance Z_p controlled through an integrating member from the s-junction. This junction contains the velocity of the valve body including an I-element for the mass, the C-element for the spring, and an effort source for perturbation forces. The force and velocity of the valve-body are connected through the transformer with pressure and volume flow of input, similar to an hydraulic cylinder. This transformer is in parallel to the modulated impedance since the pressures are equal and the volume flows additive.

Figure 4.21 shows the cross-section and a circuit symbol of an industrial piloted relief valve and Fig. 4.22 the corresponding bond graph. It comprises the pilot valve similar to Fig. 4.20 behind a

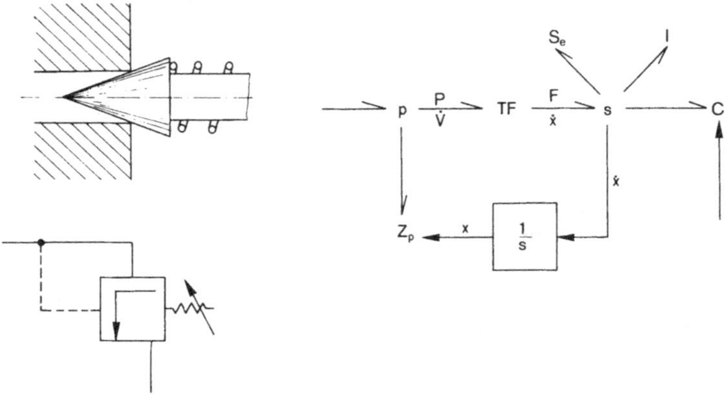

Fig. 4.20. Simple pressure relief valve; schematic and circuit symbol (left), bond graph (right).

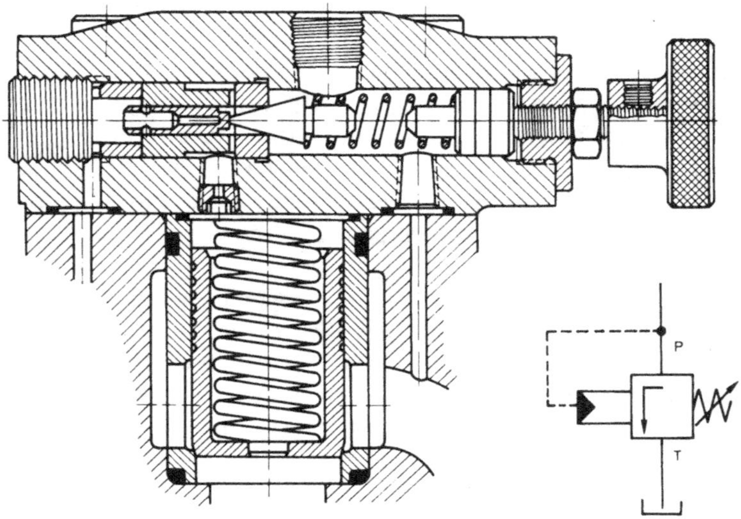

Fig. 4.21. Cross-section of an industrial piloted pressure relief valve (Abex-Denison).

restrictor Z_s and a main valve body (bottom centre of Fig. 4.21). If the pressure opens the piloting valve represented by the right half of Fig. 4.22, the corresponding volume flow produces a pressure drop at Z_s, and the pressure falls and allows the main valve body to move upwards and to open an additional flow passage. This is represented on the bond graph Fig. 4.22 by the impedance Z_h of the main valve-body modulated through the integrating member from the lower s-junction representing the velocity of the main valve-body. There act the various pressure forces trying to open and close the main valve body, as represented by the two transformers. Additionally there are the mass forces of the I-element and the momentum force (jet reaction) of the fluid modelled by the effort source. The upper s-junction carries the restrictor Z_s, whilst the top right-hand side half of Fig. 4.22 corresponds to the bond graph of Fig. 4.20 and describes the action of the pilot valve.

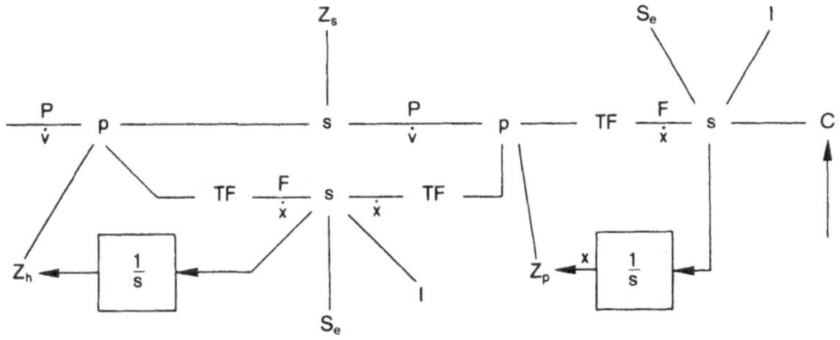

Fig. 4.22. Bond graph of the relief valve of Fig. 4.21.

Piloted pressure relief valves of this kind are frequently used for larger flows since they have very good characteristics greatly increasing the volume flow with a small increase of pressure above the opening pressure of the piloted valve (a large tangent admittance, § 2.5.3). The bond graph indicates various possible resonances due to the different C- and I-elements.

Flow control valves are the corollary to pressure relief valves since they produce a large increase of pressure drop with a small increase of volume flow over the set point (a large tangent impedance).

According to the cross-section of Fig. 4.23 the flow control valve comprises a main valve-body in series with an adjustable restrictor, called metering restrictor. The pressure drop at the metering restrictor is acting on both end faces of the main valve-body. If it exceeds the spring force, the main valve-body operates to close the flow passage and produces a pressure drop additional to the pressure drop of the metering restrictor.

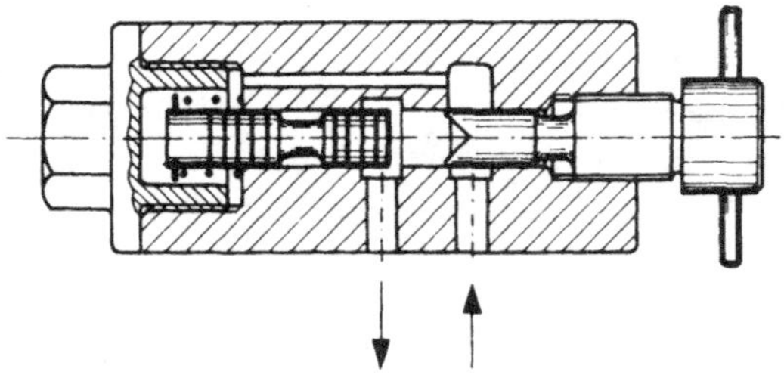

Fig. 4.23. Schematic diagram of a flow control valve.

Figure 4.24 contains the bond graph of the flow control valve. Z_h is the restriction or impedance of the main valve-body controlled

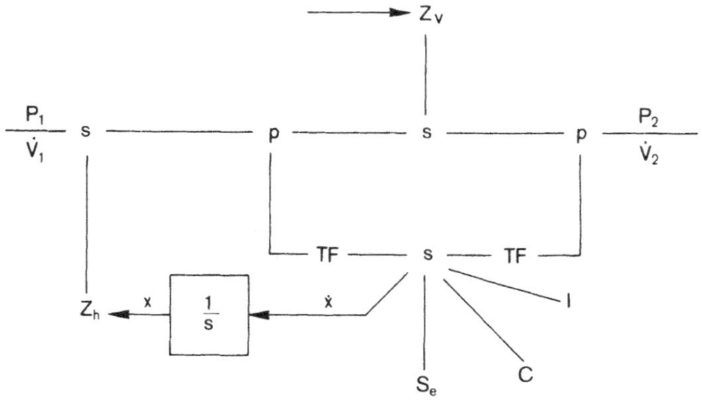

Fig. 4.24. Bond graph of the flow control valve of Fig. 4.23.

from the lower s-junction with an I-element for the mass, a C-element for the spring, and an effort source for the fluid momentum force. The transformers oppose each other corresponding to the end faces of the main valve-body, as clearly visible on the bond graph. Between the faces there is the metering restrictor Z_V in series. The fluid leaves through bond 2 with p_2 and $\dot{V}_2$. If the output pressure p_2 is zero, the p-junction (right) and the transformer disappear (§ 2.4).

There is a striking similarity between the bond graphs of Fig. 4.22 for pressure relief valves and Fig. 4.24 for flow control valves. The essential difference is that the impedance of the main valve-body is connected through a p-junction in the pressure relief valve and through an s-junction in the flow control valve, corresponding to its parallel and series connection. This is the reason for the quite different characteristics (refs. 5 and 6). The pilot valve of the pressure relief valve is replaced in the flow control valve by the output connection, in which the pressure frequently vanishes.

Bond graphs can be used to derive the principle of operation of hydraulic one-ports from the cross-section or schematic. Augmenting bond graphs and adding further elements permit detailed statical and dynamical calculations or a direct simulation of the behaviour by the ENPORT program.

CHAPTER 5

Electronics and Electromechanical Power Conversion

5.1. Electric Drive Motors

Electric applications are ideally suited for bond graphs since the concepts of resistance, inductivity, and capacity were introduced originally for electrical circuits whilst they appear now as R-, I-, and C-elements in bond graphs. On the other hand, equivalent circuits are widely used for electronics and electromechanical power conversion.

For a systematic construction of a bond graph from an electric circuit the procedure is quite similar to fluid power:

(1) To each connection of wires (point of definite voltage) a p-junction is associated. The circuit elements are added in between through s-junctions.
(2) The resulting bond graph can be simplified by eliminating all junctions with only two bonds (§ 2.4) and by assigning the voltage zero to one p-junction whence all adjacent bonds vanish.
(3) The resulting bond graph is augmented by selecting the causality and the positive power directions.

The fundamental law of electromagnetic induction appears in Fig. 5.1 as schematic (left), and as a gyrator (right). This bond graph element corresponds to the following equation with the gyrator ratio λ:

$F = \lambda i, \quad u = \lambda v.$

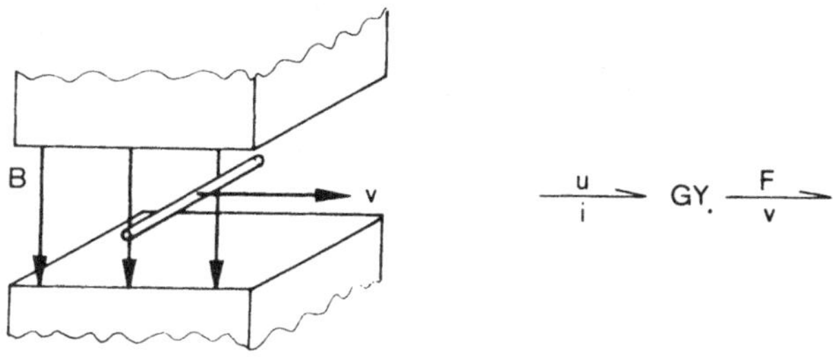

Fig. 5.1. Schematic diagram for the electromagnetic induction law (left) and bond graph (right).

The voltage is proportional to velocity, and the force proportional to the current and the constants of proportionality are equal due to power conservation. It is also proportional to the magnetic field denoted usually by B. The resistance of the cross-wire can be included by an R-element in series.

Electric d.c. motors are represented by the circuit of Fig. 5.2, where the magnetic field is approximately proportional to the field or excitation current i_e. The d.c. motor is a gyrator modulated by the

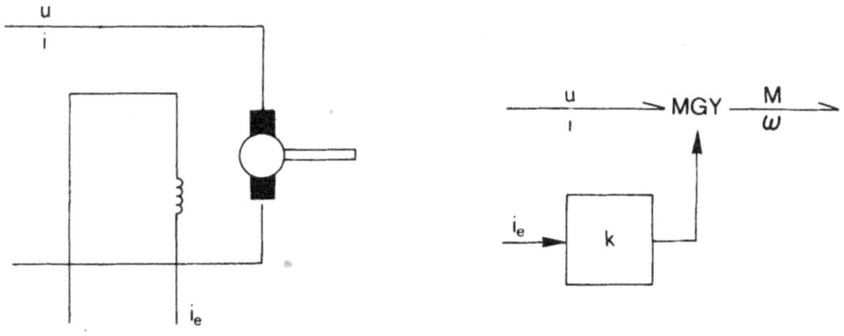

Fig. 5.2. Electric d.c. motor with separate excitation as circuit (left) and as bond graph with a gyrator modulated by the field current (right).

field current as shown by the bond graph of Fig. 5.2 (right). The field current requires no voltage in principle except for the resistance of the field windings. Consequently it enters through the activated bond without energy flow. The relation between torque, rotation frequency, voltage, and current corresponds to the fundamental equations of the gyrators:

$$M = \lambda i_a = k\, i_c\, i_a, \qquad (5\text{-}1)$$

$$u = \lambda\omega = k\, i_c\, \omega \cdot$$

The field current can be supplied from a separate source or connected in series or parallel with the armature, as shown on Fig. 5.3 as a bond graph. This corresponds to the series and shunt-excited motors. The voltage source for the electric power is a good approximation with large electrical supply systems or bus bars. The bond graphs of Fig. 5.3 include the causalities, but they are limited since the voltage source imposes an effort (the voltage) which the gyrator transforms into a flow, the rotation frequency. Strictly

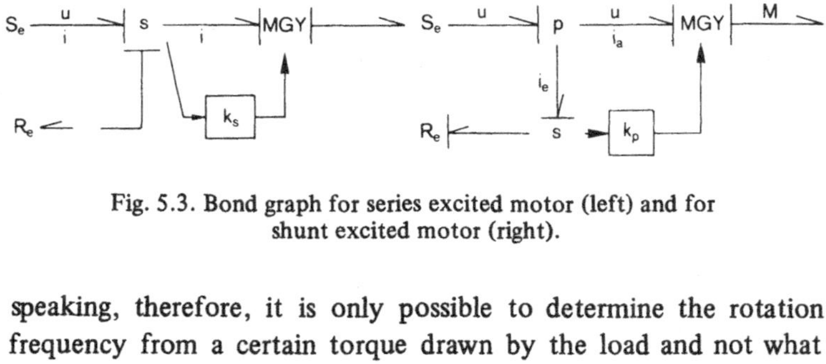

Fig. 5.3. Bond graph for series excited motor (left) and for shunt excited motor (right).

speaking, therefore, it is only possible to determine the rotation frequency from a certain torque drawn by the load and not what torque is produced with a given rotation frequency. With series-excited d.c. motors armature and field currents are equal, and this leads to

$$M = k\, i^2 \qquad u = k\, i\omega. \qquad (5\text{-}2)$$

It is possible to eliminate the common current i and to determine the torque and the absorbed electrical power as function of the rotation frequency as follows:

$$\omega^2 = \frac{u^2}{k\,M} \qquad \text{and} \quad M = \frac{u^2}{k\,\omega^2} \qquad \dot{E} = \frac{u^2}{k\,\omega}. \tag{5-3}$$

Equation (5-3) contains the known characteristics of a series-excited d.c. motor, where the absorbed power increases reciprocally to the rotation frequency and the torque reciprocally to the square or the rotation frequency. The resistance of the field windings is neglected but it could be included by an additional R-element on the s-junction, shown in Fig. 5.3 with broken lines.

With shunt-excited motors in Fig. 5.3 (right) the resistance R_e of the field winding is essentially important since it determines together with the supply voltage the field current i = u/R. It follows that

$$M = k\,i_a i_e = k i_a \frac{u}{R_e}\,. \tag{5-4}$$

The armature current cannot be eliminated from (5-4) as would be needed to determine the rotation frequency as a function of torque. This is a consequence of the causality conditions of the gyrator where with supply from a voltage source only the causality of Fig. 5.3 is admissible and not a causality with the torque acting on the motor. Inserting the field current into (5-2) one obtains instead

$$u = \frac{k}{R_e}\,u\omega \qquad \text{or} \qquad \omega = \frac{R_e}{k}. \tag{5-5}$$

The second form of (5-5) does not contain the supply voltage. Thus the shunt motor works as a pure mechanical flow source where the rotation frequency is independent of the supply voltage. This behaviour is referred to as shunt motor characteristics in electric power engineering.

The rotation frequency of the shunt motor depends only on the constant k and the resistance of the field winding as long as magnetic saturation of the exciting field is neglected. This is also true if the supply does not come from an ideal voltage source but from a voltage source with internal resistance, since the rotation frequency is independent of supply voltage.

Some electric motors have compound excitation consisting of a combination of series and shunt windings as shown in Fig. 5.4. The torque/rotation frequency characteristics can be obtained similarly as above, and the bond graph may be completed by including an R-element for the resistance of the armature winding.

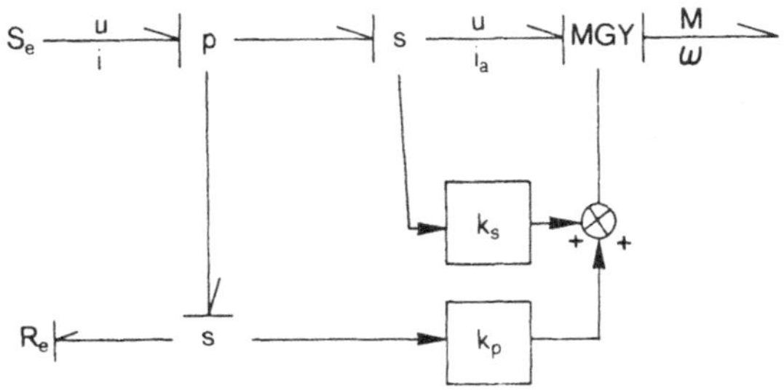

Fig. 5.4. Electric motor with combined excitation.

A more detailed description of electrical machine, including a gyrator on the mechanical side and an I-field, corresponding to the so-called transformation of Park, is contained in the important ref. 11 of Karnopp.

The electric transformers are the prototypes of the bond graph transformer denoted by TF and identical to them in the ideal case. The inclusion of losses and stray fields is treated in § 8.4.

5.2. Bond Graphs for Electric Circuits

Bond graphs for electric circuits can be constructed simply and systematically by the rules given at the beginning of § 5.1. The practical importance of bond graphs for purely electrical systems is not too great since the theory of electric circuits is well developed. However, with interdisciplinary systems as the electromechanical drives of § 5.1 and the signal transducers of § 5.3, it is also extremely useful to represent the electric parts by the same bond-graph symbols. We shall give here some examples for electric circuits.

Figure 5.5 contains a circuit where a voltage generator (shown symbolically as an electric machine) supplies first, an inductor, then

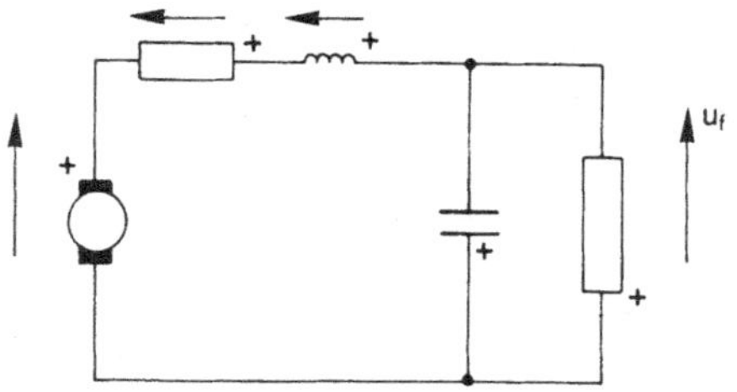

Fig. 5.5. Electric circuit with voltage generator, resistors, inductor, and capacitor.

a resistor in series, and, finally, a resistor and a capacitor in parallel. Associating a p-junction to each point with a definite voltage (i.e. to each wire connecting point), and applying the circuit elements in-between through s-junctions, the bond graph of Fig. 5.6 (top) is obtained. This figure still has a certain similarlity with the circuit Fig. 5.5.

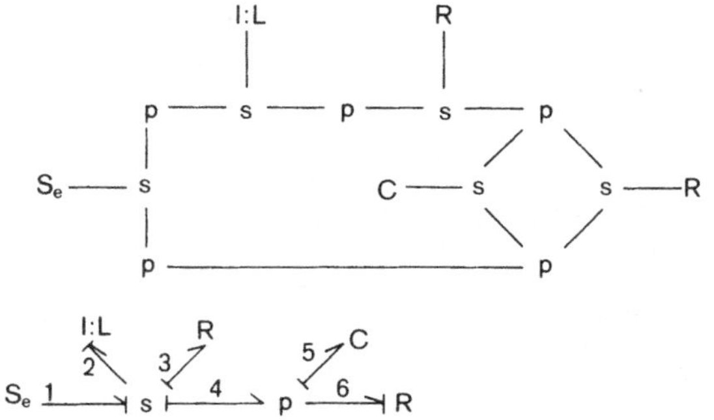

Fig. 5.6. Bond graph for the circuit of Fig. 5.5; (top) set up systematically and (bottom) simplified.

The lower part of Fig. 5.6 follows with the following simplifications:

(1) The voltage in the lower p-junctions and bonds have been set to zero and these elements vanish. This corresponds to zero electric potential in the lower connection of the circuit of Fig. 5.5.
(2) Several s- and p-junctions with only two bonds have been eliminated, i.e. replaced by a through-going bond (§2.4).
(3) The remaining two s-junctions have been replaced by a single one carrying both the I- and R-elements.
(4) The simplified bond graph has been augmented by selection of positive power directions and the causalities.

One undesirable result of the simplification is that the similarity with the circuit Fig. 5.5 has been lost. The bonds of Fig. 5.6 have been numbered consecutively (but in arbitrary order). Each one-port receives conveniently the number of its bond e.g. both R-elements are denoted by R_3 and R_6. The consecutive numbering of the bonds leads to somewhat higher numbers for the indices (instead of numbering each R-, C-, and I-element consecutively by itself) but allows a direct transfer to the ENPORT program of Chapter 7. Furthermore, a single number identifies both the bond and the adjacent element.

The power directions have been selected in Fig. 5.6 in the usual way where positive power flows out from the effort (voltage) source and into the one-ports although the real power direction reverses its sign momentarily with alternating current in C- and I-elements.

The corresponding positive voltage and currents have been refered back to Fig. 5.5 for an interesting comparison. The positive voltage is indicated there by the arrow directed upwards. With positive power flow the current of the generator is positive if it flows upwards as indicated by the plus sign. This determines at the same time the positive current for the I-and R-elements in series. The positive voltage is selected towards the left as shown by the arrows for the voltages and the plus signs for the current. These one-ports absorb energy if both current and voltage are either positive or negative. Finally, the positive current on both elements on the right must flow downwards when they absorb power.

This description shows that it is more convenient to determine first the positive power directions on bond graphs and then write them into the circuit. The bond graphs are thus also a powerful aid for the frequently complex and tedious determination of signs.

The next example is a circuit with transformer, resistor, and inductors on both sides fed through a voltage source on the left and feeding an undetermined load on the right (Fig. 5.7). The top of the

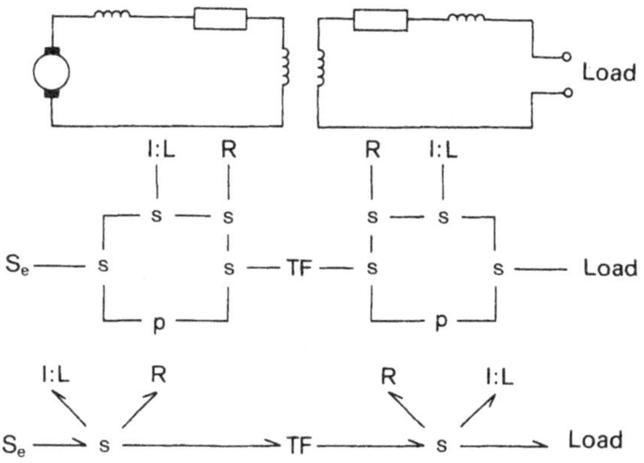

Fig. 5.7. Electric circuit with transformer; (top) as circuit, systematic and simplified bond graphs (centre and bottom).

figure contains the circuit and in the centre a bond graph where some p-junctions between the elements are already eliminated. It is still valid for any voltage in the terminals of the transformer. In particular, in the case of an insulation transformer both windings have different voltages to the ground, which attack at the lower p-junctions. Setting the voltage in both lower p-junctions equal to zero considerably simplifies the bond graph as shown on Fig. 5.7 (bottom).

Our final example is the Wheatstone bridge of Fig. 5.8 which has a strong similarity with the hydraulic control valve of § 4.3. The left part contains the circuit with a voltage generator (voltage source), the four resistors $R_1 - R_4$, and the load resistor R_L. The right part

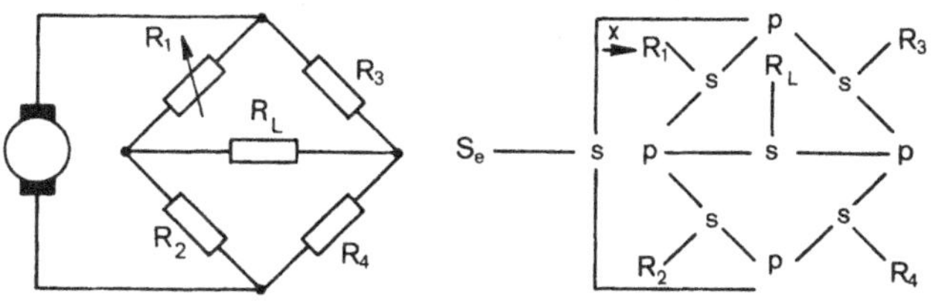

Fig. 5.8. Wheatstone bridge as circuit (left) and as bond graph (right).

contains the bond graph obtained systematically by replacing each branch point of wires with a p-junction and applying the elements between through s-junctions.

Setting the voltage in the lower p-junction of Fig. 5.8 to zero, the s-junctions of R_2 and R_4 have only two bonds each and can be eliminated. Finally, one can choose the slightly different arrangement of Fig. 5.9 obtaining a very impressive symmetrical bond graph in the form of the hydrocarbon ring of organic chemistry (ref. 2, chapter 4).

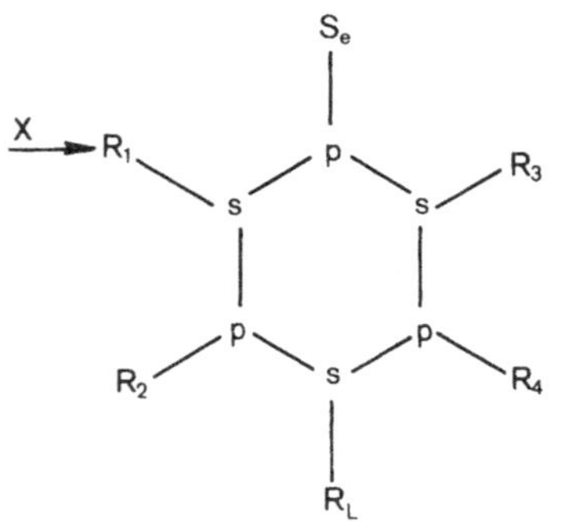

Fig. 5.9. Simplification of the bond graph of Fig. 5.8 with vanishing voltage in the lower p-junction. It has the striking form of a hydrocarbon ring.

The activated bond with the arrow on R_1 of Figs. 5.8 and 5.9 indicates that this element can be modulated by a signal, e.g. in a simple circuit of strain gauges. The more frequent push–pull circuit of strain gauges can be represented by further arrows (activated bonds).

Based on the above examples it is easy to set up bond graphs for complex electric circuits and to simplify and augment them subsequently.

5.3. Electromechanical Signal Transducers

5.3.1. *Energy storing and instantaneous transducers*

In combined electric and mechanical systems electromechanical transducers are used to connect both energy domains. With small powers the energy losses are less important than a high fidelity of signal reproduction between input and output bonds – the main requirement for signal transducers. Consequently the different losses or R-elements and the I- and C-elements carry higher efforts and flows than the power-conserving transformers and gyrators which transmit the signal. This is the main difference to the electric drives and electromechanical power converters of § 5.1 where the losses must be small perturbations.

Some components are intermediate between electromechanical power converters and signal transducers since for both a good reproduction of the signal and a good efficiency or low losses are desired.

Signal transducers can be divided into the following two classes:

(1) Transducers with instantaneous energy transmission, where the mechanical and electrical powers are equal at all times. This includes in particular the electrodynamic transducers.
(2) Transducers with energy storage, where the energy coming, e.g. from the electrical bond, is stored provisionally and subsequently withdrawn from the mechanical bond. This includes magnetic actuators and electric capacitors with moving plates (electrostatic microphones).

The operation of an electrodynamical transducer with a wire in a magnetic field has been described already with Fig. 5.1 by a gyrator. The magnetic field here is only an auxiliary to the energy conversion and the powers are equal at all times. The energy of the magnetic field is constant in time and could also be produced by a permanent magnet.

Figure 5.10 shows an electrodynamical loud-speaker as an example of an instantaneous signal transducer with a schematic (left) and a bond graph (right). The power conversion itself takes place in the gyrator; the I- and C-elements above indicate the mass of the membrane and the compliance of its suspension. The element Z_R is the radiation impedance indicating the mechanical power absorbed by the membrane or radiating as sound. I_R represents the mass of the air, moved by the membrane, called frequently apparent mass of the medium. It is sometimes included as an imaginary part (reactance) of the radiation impedance. Both depend on the signal frequency and the form of the membrane (§ 7.14 of ref. 12).

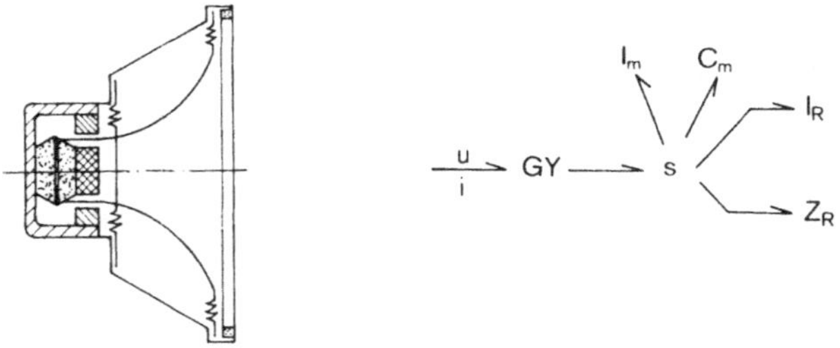

Fig. 5.10. Schematic and bond graph of an electrodynamical loudspeaker. Z_R is the radiation impedance representing the sound radiation.

It should become apparent from this example that bond graphs are useful for electroacoustic signal transducers but somewhat less for sound waves where the power is not conserved. However, in resonators with simple geometry, bond graphs for systems with distributed parameters can be used, similar to the fluid pipe of § 8.5.

5.3.2. *Capacitors with moving plates*

The capacitor with moving plates is an energy-storing signal transducer and a good example of a C-field. Figure 5.11 contains (left) a schematic and (top right) a highly idealized C-field, where the electric insulation is supposed perfect and where friction and mass of the moving plate are neglected.

The bond graph of Fig. 5.11 (bottom right) contains the insulation resistance as parallel R-element and the mass of the moving plate as I-elements. The transformer with the hydraulic impedance Z_{hydrl} modulated by the distance of the plates represents the displacement of the air between the plates, somewhat

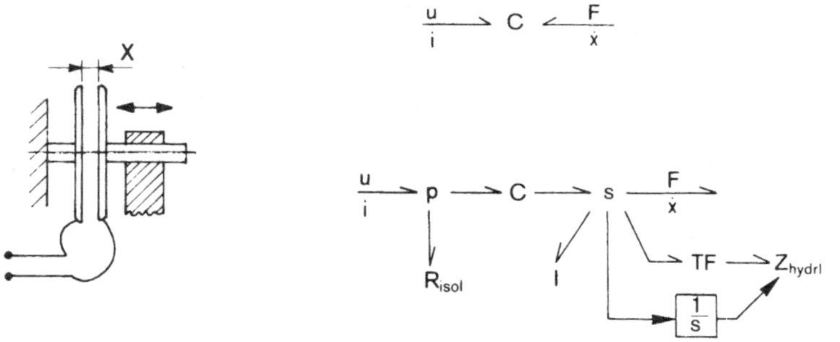

Fig. 5.11. Capacitor with moving plates; schematic (left), idealized C-field, C-field (top right), and C-field with disturbing effects (bottom right).

similar to the sqeeze oil film in the hydrostatic bearing of Fig. 4.2. Both effects are important in practice. This bond graph describes thus three engineering disciplines or energy domains (electronics, mechanics, and hydraulics) with the same set of symbols.

It should be noted that the stored energy in an ideal C-element is a definite, natural function of the electric charge and the plate separation, i.e. in general terms of both displacements. This is recognized most easily through a consideration of the power fluxes involved starting from zero charge and separation:

$$\dot{E} = u\dot{q} + F\dot{x} \qquad E = \int u(q, x)\, \dot{q} dt + F(q, x)\, \dot{x} dt \; . \tag{5-6}$$

The following substitution of the variables of integration is practical since in a C-field the efforts depend by definition only on the displacements

$$E = \int u(q, x)\, dq + F(q, x)\, dx \; . \tag{5-7}$$

The cross-coupling has the effect that the voltage depends not only on the charge but also on the plate separation. Consequently energy can be transferred from the electrical to the mechanical side and the C-field becomes applicable both for signal transduction and at least in principle also as an electric motor or generator. According to the fundamentals of electromagnetism, the relation between voltage and charge is given by the capacitance which is here a function of the plate separation x:

$$q = Cu \qquad u = \frac{q}{C} \qquad C = \epsilon \frac{A}{x} . \tag{5-8}$$

where ϵ is the induction constant and A the plate area.

A consequence of the variation of the capacitance with plate separation is the force of attraction between the plates (ref. 13),

$$F = \frac{1}{2} u^2 \frac{dc}{dx} = \frac{1}{2} \frac{\epsilon A}{x^2} u^2 = \frac{q^2}{2\epsilon A} . \tag{5-9}$$

The principle of energy conversion is to separate the plates with low charge and force, charging them up at large distance and allowing them to approach again under higher force and charge. This process requires mechanical energy for the first part but supplies more mechanical energy in the second part. Consequently there is a net production of mechanical energy and the device works as a motor. For an acceptable life of a signal transducer or motor this cyclic process must be run through a great many times – in practice many millions of times.

Figure 5.12 represents the energy conversion of a C-field by the characteristics between voltage and charge (left) and by the

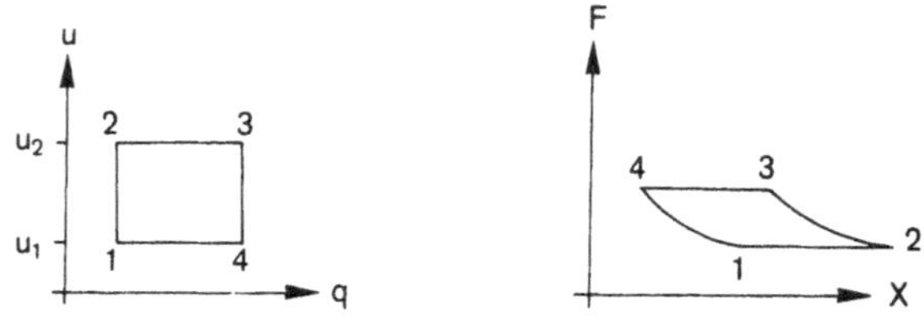

Fig. 5.12. Energy conversion by C-field with characteristics between voltage and charge (left) and between force and plate displacement (right).

characteristics between force and displacement (right). Figure 5.13 contains the bond graph for generating the characteristics of Fig. 5.12. On the left there are two voltage (or effort) sources with u_1 and u_2 and a flow source with vanishing flow. The electric bond of the C-field is coupled alternatively with these sources through the symbol td-jct for time domain junction. The td-jct is not a standard symbol of bond graphs, but useful and impressive here and in thermomechanical energy conversion.

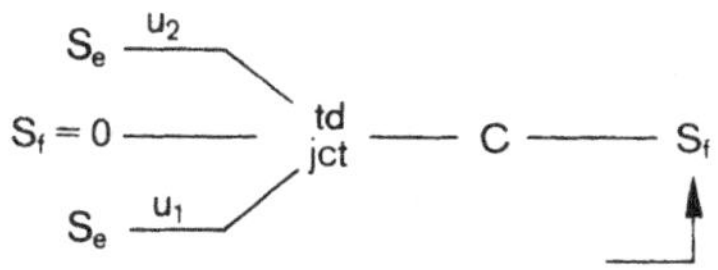

Fig. 5.13. Bond graph for generating the characteristics of Fig. 5.12.

The mechanical flow source on the right moves the plate in a way to obtain the following process.

Starting from point 1 the plate separation is increased with zero flow (connection with the flow source) whence according to (5-8) the voltage increases until u_2 is under constant charge and with decreasing capacity. At point 2 the C-field is connected with the voltage source u_2 and the separation reduces whence capacity and charge increase until point 3. Next the C-field is again connected to the flow source zero and the plate separation reduced under constant charge and force according to (5-9) until the voltage has fallen to u_1 at point 4. Finally, between point 4 and point 1 the charge is reduced with constant voltage by increasing the plate separation, thereby completing the cyclic process. The voltage source u_1 is driven in reverse by the C-field in the sense of § 2.2 and absorbs electric charge and energy.

In order to avoid losses, the connections must be established at the correct moment, i.e. if the voltage of the capacitor has just reached the value of the different effort sources. If there are errors or deviations, very high compensating currents flow shortly producing losses with the spurious resistances always present in the connecting wires.

It should be noted that for the processes of Figs. 5.12 and 5.13 some mechanical energy is first required to separate the plates against the attraction forces between point 4 through point 1 to point 2. Naturally more energy is recuperated when the plates are approaching again and the device can act as a motor. The ratio of the mechanical energy obtained to the electric energy drawn from the source u_2 is defined as work ratio, important in heat engines. The time domain junction of the electric bond corresponds to the valve arrangement of heat engines.

The interesting relations for C-fields above all follow basically from energy conservation. In fact, in a C-element the voltage is an arbitrary but single-valued function of the absorbed charge (displacement). On discharge the same voltage belongs to each value of remaining charge and consequently exactly the same energy is withdrawn. The cross-coupling between the bonds is important for C-fields since it influences, for example, the characteristics between voltage and charge by displacing the plates. This permits transferring energy from one to the other bond, and is the principle of the transducer and converters treated here. Without the cross-coupling both the electrical and mechanical part is a separate C-element conserving energy by itself.

5.3.3. *Magnetic actuators with movable iron member*

Figure 5.14 represents a magnetic actuator with movable iron member, as a schematic (left) and as a bond graph (right). This

Fig. 5.14. Magnetic actuator with movable iron member as example of a mixed IC-field.

device, a two-port with an electrical and a mechanical bond, is an example for a mixed field, a so-called IC-field for the following reasons:

(1) The electrical bond has a relation between electric voltage and time derivative of the electric current in accordance with the induction law. It is consequently an I-element with a time independent relation between electrical momentum (time integral of voltage) and the current.
(2) The mechanical bond contains a relation between the effort (force) and the displacement like a C-element.

As treated in § 8.4, the operation – and especially the cross-coupling between the bonds – result from the fact that the magnetic flux, which is really a displacement (see § 8.4) is influenced both by the electric current in the windings and by the position of the moving iron member.

The stored energy follows from the similarity to (5-8),

$$E = \int (iu + F\dot{x})\, dt\,, \tag{5-10}$$

where $u = \dot{\Phi}$ is the time derivative of the magnetic flux. Since current and force depend only on the magnetic flux and mechanical displacement, the integration variables can be changed into $d\Phi$ and dx;

$$E = \int i\,(\Phi, x)\, d\Phi + F(\Phi, x)\, dx\,. \tag{5-11}$$

The stored energy is consequently a function of both magnetic flux and mechanical displacement. With general IC-fields, energy is a function of the momenta and the displacements.

In order to identify a magnetic actuator with an IC-field it is an important condition that the magnetic flux is a single-valued function of the current and that the known hysteresis losses vanish. Thus hysteresis and eddy current losses can be simulated by additional frequency dependent R-elements.

5.4. Additional Observations on Energy-storing Transducers

5.4.1. *Maxwell relations*

In the movable plate capacitor the stored energy is, as in all C-fields, a function of the charge and the plate separation. Consequently, the voltage and the attracting force of the plates can be obtained from the principles of electromagnetism by derivation as follows:

$$u = E^{|q}_{-x} \qquad F = E^{|x}_{-q} \,. \tag{5-12}$$

Equation (5-12) uses the following simplified notation for the derivatives

$$E^{|q}_{-x} = \frac{\partial E}{\partial q} \quad x = \text{const} \qquad E^{|x}_{-q} = \frac{\partial E}{\partial x} \quad q = \text{const}\,.$$

In this notation the independent variable is written as a superscript preceded by a vertical dash and the constant variable as a subscript preceded by a horizontal dash. The latter is necessary in order to avoid confusion with normal indices.

Writing the second derivative of the energy one obtains, due to the equality of the mixed second derivatives of a function of two independent variables (so-called theorem of Schwartz if the derivatives are continuous),

$$u^{|x}_{-q} = F^{|q}_{-x} \,. \tag{5-13}$$

Equation (5-13) contains one of the important Maxwell relations. It states that the increase of the voltage due to a displacement with constant charge equals the increase of force with charge in constant position (displacement). The Maxwell relations are a consequence of the energy conservation in C-fields and are denoted sometimes as the reciprocity of Maxwell. They are the general property of C-fields and universally applicable to any energy domain, but used most frequently in thermodynamics.

In order to set up the equations for a general C-field we use the notation e_i, f_i, q_i, p_i as effort, flow, displacement, and momentum of the bond labelled i, and instead of (5-13) obtain

$$e_i = E\Big|^{q_i}_{-q_{\hat{i}}} \qquad e_i\Big|^{q_j} = e_j\Big|^{q_i}, \tag{5-14}$$

where $\hat{i}$ denotes all bonds except the one labelled i and j refers to an arbitrary bond.

Similarly, one obtains for the IC-field of Fig. 5.13 by derivation of (5-11)

$$i = E\Big|^{\Phi}_{-x} \qquad F = E\Big|^{x}_{-\Phi} \tag{5-15}$$

and the following Maxwell relation by a further mixed derivation:

$$i\Big|^{x}_{-\Phi} = F\Big|^{\Phi}_{-x} \tag{5-16}$$

The C-field can be easily changed into a causal diagram, i.e. a block diagram as shown in Fig. 5.15 in the causality where the displacements are the input variables. It is important not to forget the cross connections. Figure 5.15 (right) contains a block diagram for small deviations from the operating point, where the gains in the cross-connections, i.e. the electromechanical coupling terms, are equal by the Maxwell relation (5-13). This is a general property of all C-fields.

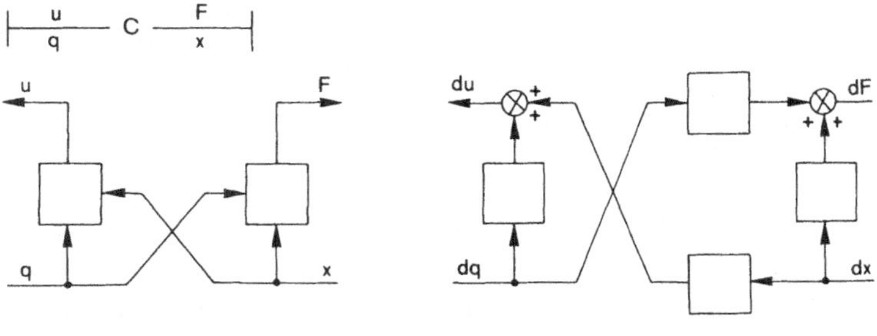

Fig. 5.15. C-field of a moving plate capacitor (top), with block diagram in general (bottom left) and for small deviations (bottom right).

The coupling terms are equal also in other causalities based on further Maxwell relations which are all obtained from the mixed second derivatives of the so-called coenergies.

5.4.2. *Energy, coenergy, and magnetic forces*

The important distinction between energy and coenergy is most clearly demonstrated on C- and IC-fields. The forces of magnetic actuators depending on the magnetic characteristics of the core material are an example.

Figure 5.16 contains a C-field for a capacitor with moving plate, (left) with an open connection represented by a flow source zero and (right) connected to a voltage source. The energy supplied through both bonds is stored in C-field due to the energy conservation. This field has been replaced in the lower part of Fig. 5.16 by a general junction and a C-element in order to show clearly the bond 3 responsible for the storage. An integration with respect to time over the powers in bond 1 and 2 determines the stored energy.

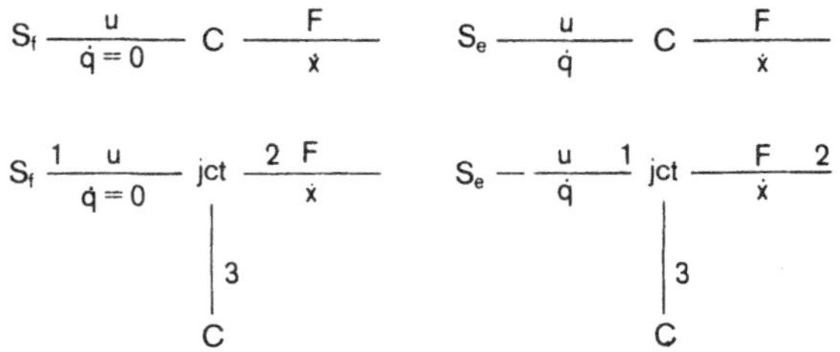

Fig. 5.16. Different representations of a C-field for illustration of the coenergy.

$$E_3 = \int u\dot{q}dt + F\dot{x}dt = \int u(q,x)dq + F(q,x)dx\,, \qquad (5\text{-}17)$$

where the substitution is again possible since all variables in C-fields depend only on both displacements.

With open electric connections no current flows and the charge is constant. The stored energy depends then only on the plate separation, and the attraction force of the plate becomes

$$F = E_3^{|x} = \frac{1}{2}\,q^2\,\frac{d}{dx}\,\frac{1}{C} = -\,\frac{1}{2}\frac{q^2}{C^2}\,\frac{dC}{dx}\,. \qquad (5\text{-}18)$$

The second and third form of (5-18) is obtained by inserting of the linear characteristics of (5-8) between voltage and charge. These forms are not valid for an arbitrary characteristic, e.g. with the so-called dielectric saturation.

The first form of (5-18) determines in general the force between the plates with open electric connection as derivative of the stored energy. The force is positive and produces an attraction with the signs of Fig. 5.16, since the capacity becomes smaller with greater plate derivation. The corresponding derivative is thus negative.

If the C-field is connected to the voltage source maintaining the voltage constant, energy is absorbed both through the mechanical and through the electrical bond

$$F dx = dE_3 - u dq \qquad F = E_{3-u}^{|x} - u q_{-u}^{|x}. \tag{5-19}$$

Figure 5.17, with the relation between charge and voltage for two different plate separations, explains (5-19). The initial characteristic a indicates an energy storage corresponding to the area 0–1–2–0.

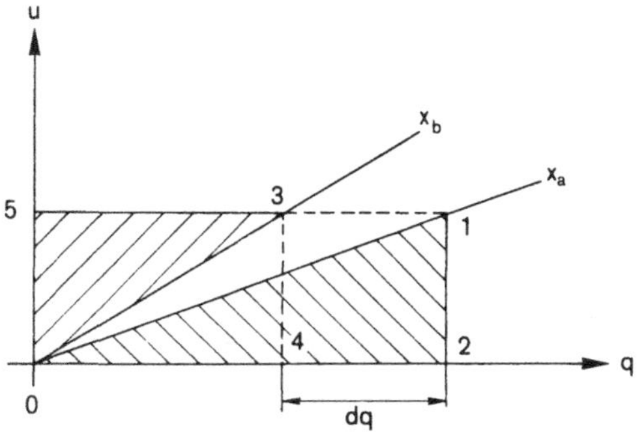

Fig. 5.17. Characteristics between voltage and charge of the capacitor for different plate separations x_a and x_b.

Displacing the plate by a small amount, characteristic b applies whence the energy 0–3–4–0 is stored. The new characteristic corresponds to a larger plate separation, and thus to a smaller capacitance producing a higher voltage with the same charge. During the displacement the energy udq, corresponding to the rectangle

1–2–3–4, has returned to the voltage source. The difference between both energies is the non-shaded triangle 0–1–3–0. It is supplied through the mechanical bond, i.e. by the attraction force of the plates.

The coenergy of the moving plate capacitor is defined as the area between the characteristic and the vertical axis of Fig. 5.17. The coenergy is thus equal to the triangle 0–1–5–0 before the displacement and equal to 0–3–5–0 thereafter. It is seen that the difference between coenergies, which are essentially associated energies is equal to the transferred real energy. The attraction force of the plate with constant voltage is thus equal to the derivative of the coenergy

$$F = -E_{CO\,-u}^{|x} \qquad E_{CO} = uq - E_3\,. \tag{5-20}$$

With a linear characteristic we have

$$q = Cu \qquad E_3 = \frac{Cu^2}{2}$$

$$F = -\frac{d}{dx}\left[u^2C - \frac{Cu^2}{2}\right] = -\frac{1}{2}u^2C_{-u}^{|x}\,. \tag{5-21}$$

The attraction force of the plates under constant voltage equals the attraction force at constant charge (open electrical connection) but only as long as the characteristic between voltage and charge is linear.

Another way of describing the attraction under constant voltage is to state that the difference in the characteristics is a triangle with an area equal to one-half the product of voltage and change of charge.

With actuating magnets the magnetic displacement (flux) ϕ and the current i replace charge and voltage, where the real energy equals the area between the x-axis and the characteristics. The attraction force with constant current corresponds to the difference area in the sense of Fig. 5.17, i.e. to the derivation of the magnetic coenergy with respect to x. The attraction force becomes with a linear characteristic

$$F = \frac{1}{2}\,i^2\,L_{-i}^{|x} = \frac{1}{2}\,i\Phi_{-i}^{|x} \tag{5-22}$$

where L is the inductance and Φ the magnetic (flux) displacement.

In the frequent case of magnetic saturation the difference area is greater, thus producing more force. Küpfmüller states (ref. 13) that the factor ½ in (5-22) becomes larger, and in the limiting case of a vertical characteristic it can reach the value of 1.

The energy as area between the characteristics and the x-axis of Fig. 5.17 is a natural function of the variable x, i.e. of the displacement of Fig. 5.17. In order to present it as function of the voltage an additional relation (in the block diagram an additional block) between voltage and displacement must be introduced. The gain of this block depends in general on the elements connected to the other bond of the C-field.

The coenergy or area between the characteristics and the y-axis is a natural function of the variable y in Fig. 5.17 of the voltage.

CHAPTER 6

Thermal and Thermodynamic Systems

6.1. Thermal Charge and Conversion Engines

Thermal energy enters easily into the picture of energy flux with the fundamental variables, effort, flow, displacement, and momentum, and consequently bond graphs are particularly applicable. The effort is the absolute temperature and the entropy flow represents the flow. It follows that entropy itself is a displacement, whilst the momentum as time integral of the temperature is not used until now. The thermal energy flux is, in the case of conduction (for convection see § 6.4),

$$\dot{E}_{th} = T\dot{S} . \tag{6-1}$$

Equation (6-1) also answers the question as to what entropy "really" is:†

(1) The entropy flow is equal to the thermal energy flux (heat flux) divided by the absolute temperature and entropy itself is its time integral.
(2) Entropy is a kind of thermal charge (refs. 8, 9) having many common properties with electrical charge. It cannot be destroyed and can be reduced in a component or element only if it is carried away by an entropy flow. On the other hand, entropy can be created by one of the processes described in § 6.3.

†In some lectures the author compares entropy to a *gray paste* as an impressive and easily memorized description.

The literature of thermodynamics describes processes conserving entropy as reversible and all entropy producing processes as irreversible. This is not very precise, since the original state of any system can always be re-established by removing the entropy. We have already seen in § 2.2 that the energy flux into an R-element is irreversible whether it is a mechanical, hydraulic, or electric resistance, and in fact all R-elements are responsible for entropy generation.

The present interpretation with entropy as principal variable of thermodynamics is also very useful in physical chemistry after Job (ref. 9), and results in great simplicity of this description. In his book (written without bond graphs) Job proposes to use the word heat to denote entropy and not thermal energy, to which the author agrees in principle. In practice, however, we shall retain the usual designations entropy and thermal energy.

In all thermal machines it is desirable to keep entropy prouuction to the minimum. This includes the desire for small mechanical and electrical losses since they also produce entropy.

Conversion engines for thermal into mechanical power, usually called heat engines, can be represented by the bond graph, Fig. 6.1 based on (6-1). The engine is idealized as not producing entropy. Hence entropy is conserved within the engine.

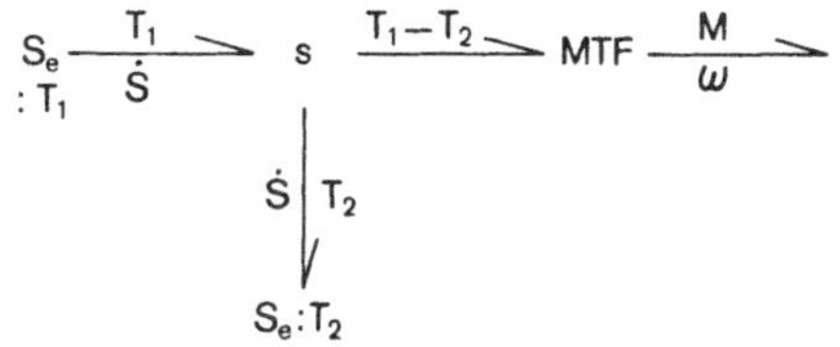

Fig. 6.1. Bond graph for conversion of thermal into mechanical energy after Carnot.

During operation the conversion engine absorbs an entropy flow from a temperature (effort) source (heat reservoir) where the temperature T_1 is independent of the entropy flow and consequently a power $T_1\dot{S}$. Due to entropy conservation the entropy flow must be disposed of in the lower temperature source with T_2, which is driven reverse in the sense of § 2.2. This requires the power $T_2\dot{S}$. From the mechanical bond a torque and a rotation frequency are taken out through the modulated transformer but no entropy flow. The available mechanical power corresponds to the difference of the incoming and outcoming thermal energy fluxes. The transformer is modulated since it does not have a definite transformation ratio but converts a thermal energy flux into a mechanical flux maintaining power.

In practice the disposed entropy flow is frequently dissipated in the cooling water without any further use. Consequently the efficiency of the conversion engine is

$$\eta = \frac{\dot{E}_{me}}{\dot{E}_1} = \frac{\dot{S}(T_1 - T_2)}{\dot{S}T_2} = \frac{T_1 - T_2}{T_1} . \qquad (6\text{-}2)$$

Equation (6-2) is the celebrated Carnot efficiency formula derived here as a direct consequence of the conservation of entropy. If the conversion engine itself produces entropy (works irreversibly), more thermal power must be spent for its disposal at the lower temperature source, and the efficiency becomes smaller.

It would be possible in principle to set up similar efficiency formulae for electric motors where the return of the electric charge takes place at appreciable voltage, but the corresponding returning electrical energy flux is normally more useful.

The conversion engine of Fig. 6-1 can also be run in reverse and thus requires mechanical power to pump entropy from the lower to the higher temperature source. This entropy pump can be used as a refrigerator or for heating of rooms where it is normally called a heat pump. In this case the entropy is taken out from a lake or a river and pumped into the room, giving in practice about three times as much heat as direct electrical heating by resistors.

Figure 6.1 does not give any information how to construct a conversion engine. This will be described now by a C-field.

6.2. Matter as C-field

The conversion of thermal into mechanical energy is possible since most substances really behave as a C-field with mechanical and thermal bonds. For instance as soon as a substance expands with entropy supply and temperature increase, there is a coupling between both bonds similar to those in an electric capacitor with movable plates.

The C-field has at the thermal side an arbitrary but single-valued relation between absorbed entropy and temperature, influenced by the mechanical bond similarly as with the electric capacitor. Figure 6.2 represents a fluid (liquid or gas) as C-field with pressure and volume as mechanical variables (right). The mechanical bond (top left) is connected with a pressure (effort) source maintaining pressure constant. The bottom (Left) part of Fig. 6.2 contains instead a flow source with vanishing flow maintaining the volume constant. The characteristics between entropy and temperature and especially their derivatives depend on the elements behind the mechanical bonds (here both sources).

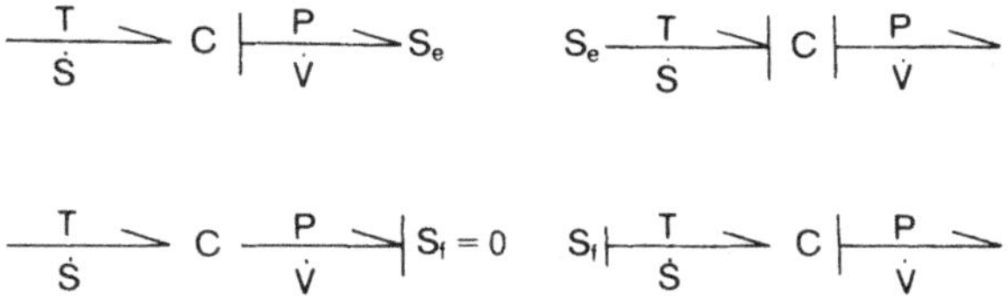

Fig. 6.2. Fluid (or any substance) as C-field with different sources on the bonds.

The entropy capacity corresponds to the more usual specific heat and constant pressure and volume as follows:

$$C_p = TS|^{T}_{-p} \qquad C_v = TS|^{T}_{-v}\,. \tag{6-3}$$

Equation (6-3) uses the same simplified notation for the derivative as § 5.4, which is especially useful in systems engineering due to its compactness. In the more conventional notation (6-3) would read

$$C_p = T\frac{\partial S}{\partial T} \quad p = \text{const} \qquad C_v = T\frac{\partial S}{\partial T} \quad v = \text{const}\,.$$

The behaviour of the mechanical bond depends on whether the thermal bond is maintained at constant temperature or at constant entropy (Fig. 6.2, right top and bottom). The inclination of the characteristic between volume and pressure is the compressibility with constant temperature or constant entropy (isothermal or adiabatic compressibility). The ratio of these compressibilities is the same as the ratio between the entropy capacities (specific heat) at constant pressure and volume (ref. 19, p. 51). This is easily demonstrated with block diagrams.

It should be particularly noted that the above relations are general properties of all C-fields. The only conditions for their application to real devices are that there are no losses or R-elements in parallel or series and the sealing and insulation are perfect.

Figure 6.3 represents the characteristics between temperature and entropy of two substances at constant pressure – one the so-called ideal gas and the other a mixture of water and water vapour under constant pressure. With this mixture the temperature remains constant over a wide range of entropy in which the evaporation takes place.

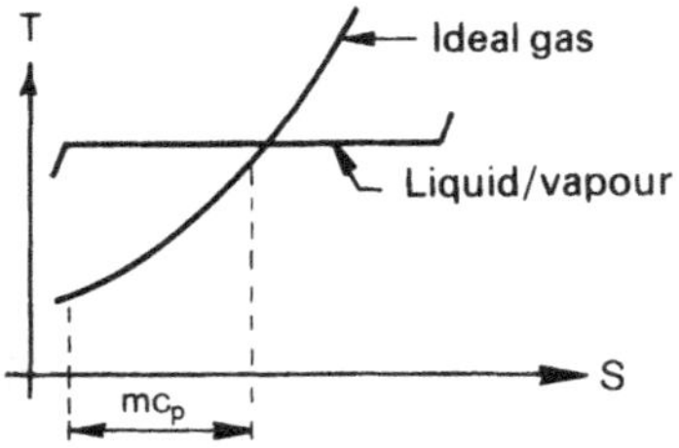

Fig. 6.3. Characteristics between temperature and entropy under constant pressure for two substances.

The volume increases considerably during evaporation and the temperature is a function of pressure. The evaporation entropy at 1 bar is for many one-atomic substances (ref. 9, p. 134) equal to 100 J/K per mole (rule of Richard). For melting solids about 10 J/K per mole are required (rule of Pictet/Trouton).

Within the range of evaporation the mixture of liquid and vapour behaves as a C-element with infinite capacitance.

Figure 6.4 contains a bond graph of a conversion engine for thermal and mechanical energy. It works between two temperature sources with T_1 and T_2, the latter for disposing of the entropy. The

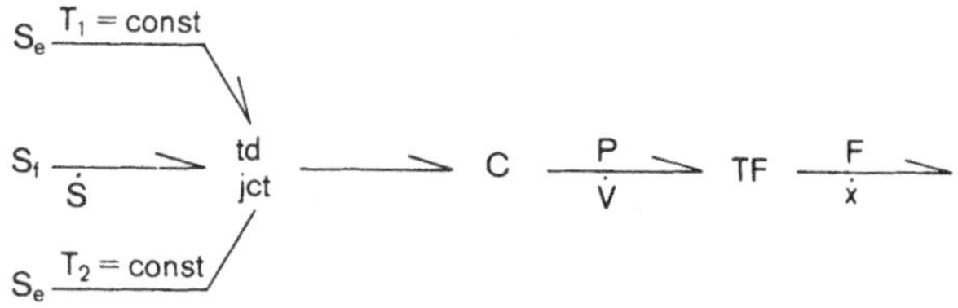

Fig. 6.4. Bond graph with C-field and time domain junction for a conversion engine.

time junction (td-jct) already introduced in § 5.3 allows connection of the C-field alternatively with both temperature sources or to separate it as represented by connection to the zero flow source. The operation of the time junction is taken care of by the valve mechanism in a real heat engine. The mechanical bond of the C-field contains a transformer producing force and velocity from pressure and volume flow and its transforming ratio is the piston area.

Figure 6.5 contains the thermal (left) and the mechanical characteristics (right) of the conversion engine with a C-field. Starting from point 1 the temperature decreases whilst the volume increases at zero entropy flow or constant entropy from T_1 to T_2.

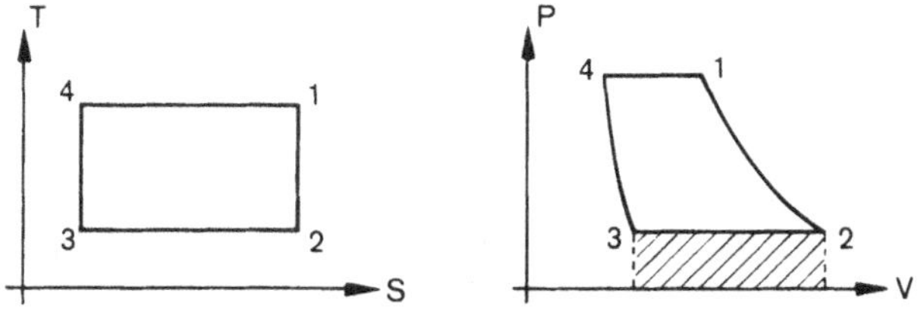

Fig. 6.5. Characteristic between temperature and entropy (left) and between pressure and volume (right) of C-field of Fig. 6.4.

Beginning at point 2 the working fluid is connected to the temperature source T_2 which absorbs entropy under constant pressure during volume reduction. The next phase is an increase of temperature and of pressure during volume reduction and vanishing entropy flow. During the final phase between point 4 and point 1, the C-field absorbs entropy from the temperature source T_1 whilst the volume increases.

The ratio of the net mechanical energy to the gross mechanical energy released during volume increase is the work ratio, essentially the same as found for the movable plate electric capacitor in § 5.3. It should be as near as possible to 1.

An exact operation of the time junction or valve mechanism is important for loss-free operation because the connections of the thermal bonds with the temperature sources must be established precisely when the working fluid has reached the corresponding temperature. Otherwise there would be entropy generation by heat conduction between different temperatures (§ 6.3).

The characteristics of Fig. 6.5 corresponds approximately to a mixture of water and vapour or generally of two phases. With ideal gases the rectangle of the mechanical characteristics is much more narrow and the work ratio correspondingly low.

The conversion process of Figs. 6.4 and 6.5 between two temperature sources is the so-called Carnot cycle. It consists at the thermal side of two characteristics with constant temperatures and two characteristics with constant entropy denoted usually as isothermals and adiabatics. The optimum efficiency of (6-2) can be obtained with it. However, this is not applied in practice due to (a) the small work ratio with ideal gases, (b) the condensation of vapour (leading to erosion of turbines by water drops), and (c) to the difficult valving conditions. Furthermore, thermal power is not usually available from a temperature source but rather in a gas stream (combustion gases or cooling gases of nuclear reactors). Such a stream cools down if entropy flow is withdrawn. Consequently other working cycles with different thermal and mechanical characteristics are used in practice, like the Joule, Otto, Diesel, and Stirling cycles.

In conclusion, it should be noted that the greater part of conventional thermodynamics is concerned with thermal C-fields. An important subject is the equilibrium with different elements on the thermal and mechanical bonds. For the calculation of the equilibrium new variables like enthalpy, free energy, and free entropy are introduced, which are the coenergies dealt with in § 5.3.

The entropy is the co-energy of the mechanical bond, the (Helmholtz) free energy the coenergy of the thermal bond, and the free enthalpy (Gibbs free energy) the double coenergy on both bonds.

Quite different is the thermodynamics of flow processes also called irreversible thermodynamics. This deals essentially with thermoconduction and R-fields and will be briefly treated in § 8.3.

6.3. Generation of Thermal Charge

The thermal charge or entropy can be produced "out of nothing" by the following (irreversible) processes:

(1) Due to frictions of any kind – mechanical, hydraulic, electric, or chemical (in chemical reactions), i.e. each time where an R-element appears in a bond graph.
(2) By thermal energy flux (heat conduction) with appreciable temperature drop.
(3) By mixing of two fluids with different temperatures or of different chemical nature.

The thermal charge or entropy is a positive definite variable, and during its generation no negative charge remains as in electronics. Furthermore it is not the starting point of a force field as is the electric charge.

Since entropy is generally undesirable and energy must be expended for its disposal, machines and engineering systems should be designed for minimum entropy production.

Thermodynamics is essentially an extension of mechanics and electronics by including the thermal energy. With this inclusion energy is universally conserved, and losses produce a corresponding thermal energy flux by the R-elements of bond graphs. This universal energy conservation (including thermal power) is known as the first law of thermodynamics.

If thermal effects are included in the model the R-elements of bond graphs must be replaced by a new element, the RS-field. It has at least a mechanical or electric bond and a thermal bond from which the loss power leaves in the form of newly generated entropy flow and temperature.

The bond graph of Fig. 6.6 contains (left) an RS-field producing entropy flow and temperature from torque and rotation frequency, or any other mechanical or electrical variables. The following modulated transformer represents the conduction of the entropy flow under appreciable temperature drop as described further below.

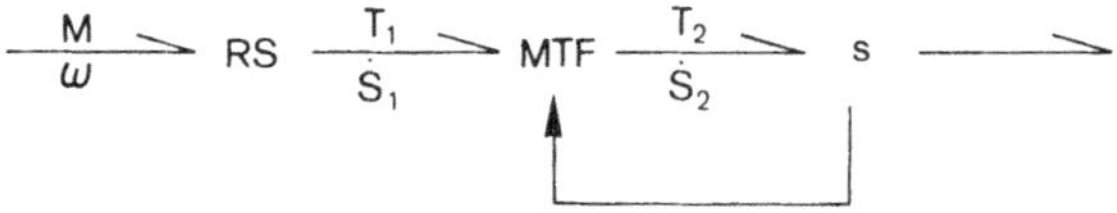

Fig. 6.6. RS-field with subsequent modulated transformer for entropy generation by friction.

An RS-field is power conserving and always works in the direction in which thermal energy is generated. The mechanical characteristics are frequently influenced by the thermal variables, e.g. the temperature dependence of the friction in oil (viscous friction) or of electrical resistance. Each mechanical or electrical flow through an R-element produces a thermal energy flux composed of entropy flow and temperature depending on the elements terminating the thermal bonds.

Referring now particularly to heat conduction or entropy flow with appreciable temperature drop additional entropy flow is generated. It can be represented as follows:

(1) by a modulated transformer;
(2) by an RS-field similar to the mechanical and electrical losses.

Referring to the entropy generation by a modulated transformer (Fig. 6.6, right) the energy flux (thermal power) is conserved and the entropy flow increases corresponding to the temperature drop

$$\dot{E}_{th} = T_1\dot{S}_1 = T_2\dot{S}_2 \qquad \dot{S}_2 = \frac{T_1\dot{S}_1}{T_2} \tag{6-4}$$

The transforming ratio is not fixed but depends on the effort and flow variables, as indicated in Fig. 6.6 (right) by the activated bond from the s-junction modulating the transformer.

For entropy generation by RS-fields it is observed that the entropy flow with appreciable temperature drop generates losses similar to the electrical or mechanical flow processes. Consequently a corresponding entropy flow appears in the bond with the lower temperature. Figure 6.7 shows a schematic of a heat conducting bar with the temperatures T_1 and T_2 at the end faces (left) and the RS-field for generation of the entropy (right). The loss power corresponds to the temperature drop and to the entering entropy flow and is thus brought to the RS-field from an s-junction. The

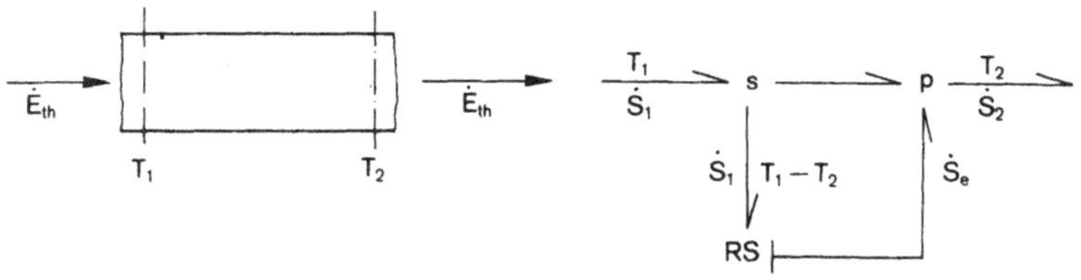

Fig. 6.7. Entropy generation by heat conduction; schematic (left) and bond graph with RS-field (right).

newly generated entropy flow is fed into bond 2 through the p-junction. The equation for the entropy generation is

$$\dot{S}_e T_2 = \dot{S}_1 (T_1 - T_2) \qquad \dot{S}_2 = \dot{S}_1 + \dot{S}_e \tag{6-5}$$

where $\dot{S}_e$ is the generated entropy flow in accordance with (6-4).

Figure 6.7 also contains the half-arrows for the positive energy fluxes, indicating that power enters the RS-field and leaves with the newly generated entropy flow. This flow cannot reverse its direction even if the direction of temperature drop should change ($T_2 > T_1$). In this case the new entropy flow would enter through a p-junction into bond 1. Finally, the usual causality is indicated on Fig. 6.7, where the temperature is responsible for the new entropy flow.

More complex RS-fields with two ingoing bonds (e.g. electric and thermal) and a bond for the new entropy flow will be described in § 8.3. With such fields the direction of flow can reverse in one of the input bonds as long as some loss remains and some new entropy is generated.

It should be noted that the torque and rotation frequency in the former Fig. 6.6 can be replaced by any other effort and flow variable. The RS-field in this figure is followed by a modulated transformer describing the following heat conduction and entropy generation as already mentioned.

It would be possible in principle to consider the temperature T_1 as infinite (Ref. 8) whence the entire entropy flow is generated in the transformer. This might appear artificial but is a good model for the overheating of friction brakes and electrical resistors with insufficient entropy removal or cooling.

For comparison with the more conventional description we give here the formula for heat conduction where the conductivity k depends generally on the temperature T and where A and L denote the cross-section and the length of the bar:

$$\dot{E}_{th} = k\frac{A}{L}\ (T_1 - T_2). \tag{6-6}$$

It follows for the entropy flows

$$\dot{S}_1 = \frac{\ddot{E}_{th}}{T_1} = k\frac{A}{L}\ \frac{T_1 - T_2}{T_1}\ . \tag{6-7}$$

The ratio of the entropy flows, i.e. the transforming ratio of the modulated transformer of Fig. 6.8 is obtained by combining (6-4), (6-5), (6-6), and (6-7):

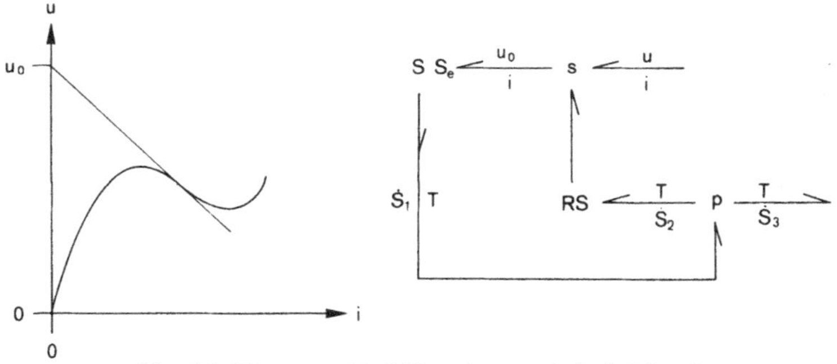

Fig. 6.8. Element with falling characteristic (left) and corresponding bond graph (right).

$$\frac{\dot{S}_2}{\dot{S}_1} = 1 + \dot{S}_2 \frac{L}{kA} \, .$$

Equation (6-8) shows how the transforming ratio depends on the outgoing entropy flow $\dot{S}_2$ as indicated by the modulation from the s-junction on Fig. 6.8.

The transforming ratio is always larger than 1 as long as entropy flows from left to right. Within reversed temperature drop the entropy flow also reverses whence the transforming ratio becomes smaller than 1 and the entropy input flow on the right is less than the outgoing flow on the left. Consequently the modulated transformer always acts to produce new entropy.

If entropy is once generated it cannot be destroyed but only transported away and deposited elsewhere. In this sense any entropy producing process is irreversible.

The modulated transformer of Fig. 6.6 correctly represents the experience that thermal energy runs only from high to low temperature and never the reverse. On the other hand, it is possible to pump entropy from low to high temperature expenditure of mechanical power (§ 6.1).

Some characteristics in electronics and mechanics have a falling slope, where the voltage drop decreases with increasing current (Fig. 6.8). Such characteristics are described by a voltage source with negative internal resistance. With the characteristics of Fig. 6.8 the voltage source is driven in reverse and absorbs electrical power which is returned partly to the electric circuit by the negative internal resistance. The voltage source thus becomes a new bond graph element, the double source S5. It supplies thermal power into the RS-field of the internal resistance which produces electrical power from an entropy flow. This anomaly is not in contradiction with the conservation (indestructibility) of entropy since the double source incorporated in the same real component always produces a larger entropy flow. It is given to the environment through the p-junction.

Alternatively one can state with reference to § 2.5 that the secant impedance responsible for entropy generation is always larger than zero. The negative tangent impedance is artifical for representing the falling characteristics, but in reality it is always associated with the reverse-driven double source.

The negative internal resistance can lead to oscillations if electric power is fed through a suitable source as frequently experienced in electronics. Similar oscillations can be excited in mechanics if the friction force decreases with increasing velocity.

6.3.1. *Fundamental laws of thermodynamics*

In conclusion we shall cite here the fundamental laws of thermodynamics although they are implied partly from the above.

Zeroth law If the temperature of a first piece of matter is equal to a second, the temperature of the second to a third, then the temperatures of the first and the third pieces of matter are equal. In other words this means that temperature is a universal transitive quantity.

First law Energy is universally conserved. Consequently energy fluxes add up to zero except for possible storages.

Second law Entropy cannot be destroyed but only generated. All processes in R-elements move in the direction generating entropy.

Third law At absolute zero temperature all bodies have zero entropy contents (theorem of Nernst, 1908).

The zeroth law was formulated later but is stated first because of its fundamental statement about the nature of temperature.

It results from our treatment that entropy generation by friction and heat conduction can be displayed very effectively by bond graphs. The entropy generating mixing of substances is less amenable to bond graph treatment (ref. 9).

6.4. Thermal Energy Conduction and Convection

The treatment of thermal energy in the present chapter and particularly (6-1) refers to heat conduction, i.e. entropy flow without movement of matter or substances. On the other hand, convection is the transport of energy and entropy by moving substances, in practice frequently warm water or steam-pipes. Since a substance is a C-field (§ 6.3) convection is in principle a transport of C-fields.

The difference between convection and conduction is not limited to thermal energy but a general question of systems engineering. As an example of the convection of electrical energy (Fig. 6.9) shows a large number of electric capacitors on a conveyor belt between a charging and a discharging station. Defining the transport voltage as

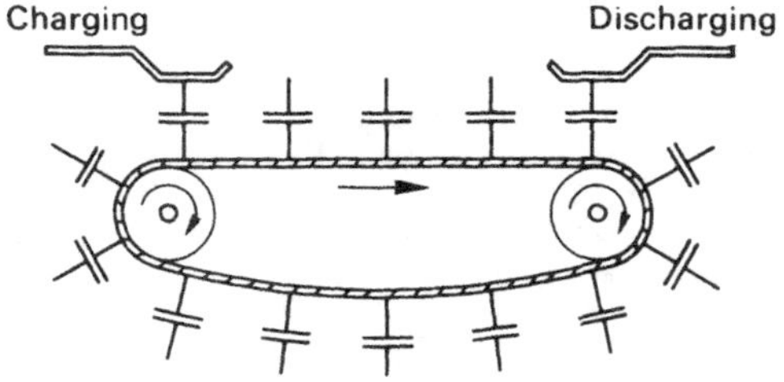

Fig. 6.9. Conveyor belt with capacitors as example of the convection of electrical energy.

the voltage reached by the capacitors at the end of the charging process and the current as their charge multiplied by the transport frequency (number of capacitors per second), the transported power is not equal to the product of voltage and current. Rather it is necessary to introduce a form factor α variable between zero and one:

$$\dot{E} = \alpha ui = \alpha T\dot{S}. \tag{6-9}$$

The second form of (6-9) refers to the convection of thermal energy.

The form factor α depends on the characteristic of the C-element, in the electrical case on the relation between voltage and charge. If the voltage is independent of the charge, the form factor becomes equal to 1, similar to the liquid and vapour mixtures of Fig. 6.3.

Normal electric capacitors have a linear relation between voltage and charge according to (5-8) and the form factor is equal to one-half. This results from the fact that the voltage of the capacitor decreases linearly during discharge and the stored energy is only one-half of the product of voltage and charge (5-21), corresponding to the area below the characteristics in Fig. 5.17. The form factor is also related to the question of the return transport of the discharged capacitors and to the zero point of entropy and of electrical charge.

Let us note at this point that the capacitors can be stowed so tightly on the conveyor belt that they exert an appreciable force upon each other. Due to this force they transport additional mechanical power which corresponds to the hydrostatic power carried in pressure and volume flow of fluid pipes (see below).

The capacitors should not simply be connected to a fixed voltage source for charging, but charged through a continuously increasing modulated voltage source. Otherwise the initial connection with the discharged capacitor results in high currents, sparks, and general losses. Similarly for discharging, the capacitor must be connected to a modulated voltage source. This source is driven in reverse in the sense of § 3.4, and its voltage always a little smaller than the voltage of the capacitor. The velocity of discharge results from the rate of change of the voltage and is influenced by spurious resistances (R-elements).

The conveyor belt with electric capacitors or with hydropneumatic accumulators for fluid energy is not a practical proposition, but only used here as an example to illustrate the relations between the various quantities during convection. Nevertheless, it has a certain similarity with the electrostatic van de Graaff generator and could be approximated by an installation supplying accumulators to a fleet of electric vehicles through a conveyor belt.

The above considerations about C-fields and about convection lead to the conclusion that thermodynamics is really 70% systems engineering and only 30% of thermics, i.e. the special properties of thermal charge or entropy described in § 6.3.

As with electric power, certain precautions have to be taken for the convection of thermal energy especially during charge and discharge in order to avoid losses – in the thermal case entropy generation. The effort variable; here the temperature depends on the absorbed entropy just as with electric capacitors. Modulated temperature sources do not exist in practice, although they are frequently postulated in thermodynamic textbooks as a number of heat reservoirs. In practice the charge and discharge take place in heat exchangers.

With internal combustion engines of the Otto or Diesel cycle or gas-turbines with internal combustion, the chemical energy of combustion is liberated. This is represented by the addition of a thermal energy flux.

Energy transport through flowing substances under pressure incorporates really two parts similar to the conveyor belt with tightly squeezed capacitors:

(1) An external or hydrostatic power transport given as a product of pressure and volume flow.
(2) The transport of internal energy by C-fields by convection.

Both parts are suitably represented in a bond graph by the separate bonds of Fig. 6.10. On top there is the transport of C-fields with its relation between entropy flow, temperature, and transported energy – the so-called internal energy (6-9). The lower bond contains the flux of mechanical of hydrostatic energy as pressure and volume flow. Just as with the tight capacitors it is not carried by the fluid itself and does not change its state.

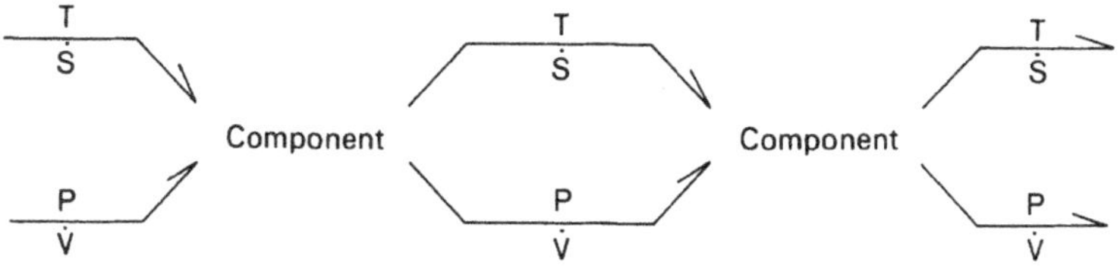

Fig. 6.10. Bond graph for entropy transport by flowing substances (so-called vector bond graph).

In general all components traversed by a fluid change the values of entropy flow, temperature, pressure, and volume flow in accordance with the characteristics of the C-field representing the fluid. Consequently the distribution of the energy flux between both bonds is changed. There are the following special components:

(1) Adiabatic turbines and compressors that add or withdraw

mechanical power without adding or generating entropy in the ideal case.

(2) Heat exchangers with negligible pressure drop that introduce or withdraw an entropy flow at a given constant pressure.

In many practical conversion engines the working fluid as the C-field is not located in a definite place but circulates through the engine, and each change of state takes place during the flow through a component. Figure 6.11 shows a bond graph of a conversion engine with an adiabatic turbine and compressor and with heat exchangers for introduction and withdrawal of the entropy flow according to the so-called Joule cycle. The simple bond between turbine and compressor shows the power needed to drive the latter. The work ratio (§§ 5.4 and 6.3) is approximately equal to the ratio of turbine power minus compressor power divided by the turbine power. It indicates which fraction of the turbine power is needed for the drive of the compressor. With a low work ratio the turbine power is much larger than the net output power and the unavoidable losses of turbine and compressor have much effect.

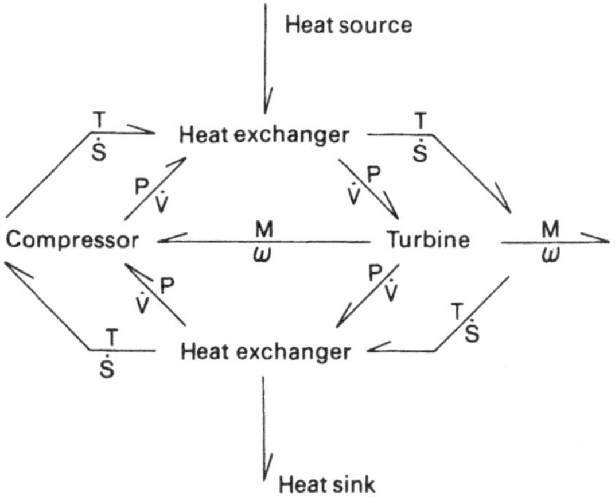

Fig. 6.11. Bond graph for a conversion engine with flowing working substance.

Another characteristic quantity of conversion engines with circulating working fluid is the specific circulation, i.e. the mass flow per output power (kg per kWsec). The higher the specific circulation the more mechanical power must be expended for driving it around.

It should be noted that even heat exchangers under constant pressure produce mechanical or hydraulic power from the incoming thermal power flux. This is the case if the working fluid expands during entropy addition (heating), as is particularly the case during evaporation of water. The hydraulic power can be transformed into mechanical power without change of state of the working fluid. With steam-engines this is the so-called full-pressure engine, but the expansion of steam has been introduced already by James Watt. It allows the transformation of the internal energy of the working substance partly into mechanical power by discharging the C-field.

Watt has observed that steam shot at "gale force" from the cylinder into the condenser and introduced expansion in order to take advantage of this force. Reduction of heat and coal consumption was already important during the time of Watt (ref. 15).

Figure 6.12 shows the conversion engine of Fig. 6.11 as circuit with symbols from chemical engineering. The circulation of a working fluid through a component can be made by the following two methods:

(1) By the forced convection where the working substance is driven by appropriate pumps.
(2) By the free convection in vertical pipes.

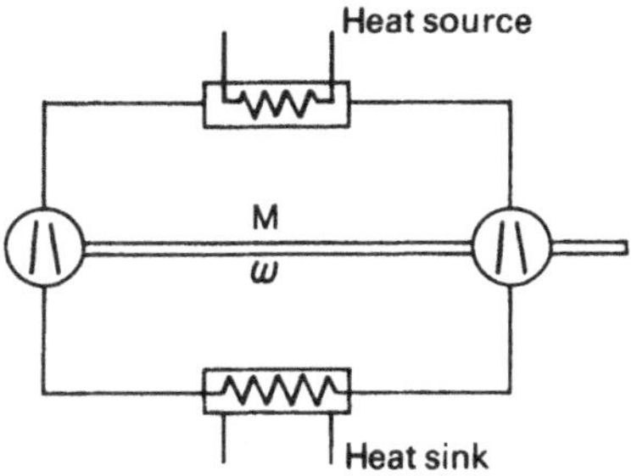

Fig. 6.12. Circuit of the conversion engine of Fig. 6.11 with symbols from chemical engineering.

In free convection the gas expands due to entropy supply or combustion. Consequently it experiences a lower force of gravity than the cold external air. This produces a pressure difference driving the circulation, which is an example of the generation of hydraulic power in a heat exchanger. Here the flow is essentially driven by the change of density of the fluid with temperature.

In order to design conversion engines with circulating substance for different working cycles it is necessary to combine supply of entropy and pressure change depending on the characteristic of the C-field. This requires turbines and compressors with simultaneous supply of entropy and is rarely used. However, turbines with reheat and compressors with interstage cooling are an approximation to these components.

CHAPTER 7

Bond Graphs and Electronic Data Processing

7.1. Fundamentals of Systems Simulation

So far we have used bond graphs essentially for the study of interdisciplinary engineering systems and for the uniform description of mechanical, electrical, and thermal effects. Apart from this they can be used to obtain, almost automatically, computer programs for the simulation of engineering systems with very little routine work for the engineer.

Within the scope of this book it is not possible to give an exhaustive treatment of the direct computer programming based on bond graphs. It will be sufficient to outline the programming method, whilst for further details, ref. 2 – especially the "User's guide to ENPORT" – should be consulted. Reference 4 contains a survey of computer programming questions.

The fundamental purpose of computer simulation from a bond graph is to determine the behaviour of certain output variables of an engineering system for given variables. The input variables are frequently step functions, or functions that increase suddenly from zero to one at the zero time.

The ENPORT simulation program allows the feeding of the structure, and numerical values of an engineering system directly from the bond graph into the computer as will be shown by an example. Just as with computer programs it is based on the state variables, the displacements, and momenta introduced with Paynter's tetragon (Fig. 2.14).

The different displacements and momenta are usually combined into a column vector **x**, the so-called vector of state (we print a variable in bold if it represents a vector or a matrix with several components). The input and output variables are combined into the vectors **u** and **y** and the state space equation are written with the general functions **F** and **G**.

$$\dot{\mathbf{x}} = \mathbf{F}(\mathbf{x}, \mathbf{y}),$$
$$\dot{\mathbf{y}} = \mathbf{G}(\mathbf{x}, \mathbf{u}). \qquad (7\text{-}1)$$

Equation (7-1) is a short notation for as many scalar equations as the **x** and **y** vectors have components. In the linear case, these equations simplify to the standard form of state space equations with the matrices **A, B, C,** and **D,** (ref. 16).

$$\dot{\mathbf{x}} = \mathbf{Ax} + \mathbf{Bu},$$
$$\dot{\mathbf{y}} = \mathbf{Cx} + \mathbf{Du}. \qquad (7\text{-}2)$$

The essential purpose of the ENPORT program is to determine the state space equations automatically from a bond graph.

Equations (7-2) are all written in integral causality, since each component of the state vector is obtained by integration with respect to time of the corresponding right-hand side of the equation. These equations are thus a system of coupled differential equations of the first order.

Whilst with almost all computer programs diagrams are used in one form or another, the different programs can be classified after Karnopp (ref. 4, § 3.6) as follows:

(1) Programs where diagrams have only minor importance.
(2) Programs based on causal diagrams.
(3) Programs based on acausal diagrams.

With the programs of the first kind the state space equations must be entered in the form of (7-1) or (7-2). The vector functions are not available in closed form but only as subroutines which must be established frequently for the case in point.

Programs based on causal diagrams are frequently used. They have the important drawback that it is difficult to set up causal diagrams (block diagrams of control engineering) for systems of a certain size with many feedback connections. Causality must be selected by hand, i.e. by the engineer himself. Only bond graphs allow determination of the structure of the system first and then to select systematically the causalities, which is a great help. Furthermore bond graphs allow changing of the causalities as will be frequently necessary on trial runs of the computer, e.g. to make a more detailed model for higher accuracy or a more simple model for reduced computation time.

With programs based on acausal diagrams, the form of the state space equations and the causalities are set up by the computer. This has been done for some systems especially in electronics directly from the equivalent circuit. The ENPORT program also belongs in this class since it starts from a bond graph without causality, i.e. an acausal diagram. Only the directions of positive energy fluxes need be selected by the engineer himself. As a preparation to this program the formulation of the state space equations by hand will be described briefly.

Analogue computers have been somewhat eclipsed by digital computers in recent years, but they remain extremely useful for certain computational tasks especially in industry. They can be programmed directly from a bond graph as described in detail by Brouer in ref. 17. This author uses an example of an electrohydraulic control system for a conveyor belt transporting objects through an oven with the requirement to control the temperature on leaving the oven. The analogue circuit of the system is obtained directly from a fully augmented bond graph avoiding the already complex block diagram. It appears that the direct programming of analogue computers from bond graphs is frequently both possible and efficient.

7.2. Manual Formulation of State Space Equations

The formulation of state space equations from a fully augmented bond graph with integral causality comprises the following three steps:

(1) Selection of the input variables, the state variables, and the variables of the coenergy.
(2) Formulation of a temporary system of equations.
(3) Reduction of the system of equations to the standard form of state space equations (7-1) and (7-2).

State variables are the natural variables of each C- and I- element as described in § 5.4. In accordance with the integral causality this is the momentum for the I-element and the displacement for the C-element. Each bond is labelled conveniently with the time derivative of the momentum instead of the flow during the formulation process. To this have to be added the natural variables of the coenergy (coenergy variables) which are the flows and the efforts. The flow is put by each I-element into a static relation with the momentum and the effort by each C-element into static (time independent) relation with the displacement.

Subsequently the system of equations if formulated with each I-element containing the above relation between momentum and flow and each C-element the relation between displacement and effort. Reducing the system of equations (step 3) allows the elimination of the variables of coenergy, i.e. efforts and flows. Details are contained in ref. 2, chapter 5, including the case of bond graphs containing unavoidable derivative causalities.

7.3. The ENPORT Program

The ENPORT program starts from a simple, non-augmented bond graph and first gives its structure by a special serial code into the computer. For this purpose the bonds are numbered consecutively and the symbols of the adjacent elements are entered into punched cards. The computer requests subsequently the values of the parameters and the form and values of the input variables. The result of the computation is information about characteristic values stability, and of the interesting output variables usually plotted automatically as function of time.

Figure 7.1 contains as an example the cross-section through a muffler, (an outlet (right) is missing, but could be represented by an R-element) and Fig. 7.2 a provisional bond graph. It indicates that resistance and inertia of the three cross-pipes are to be included, but of the annular chambers only the capacitance, and of the longitudinal pipe both the inertia and the capacitances. This model is already quite complex but does not yet include the resistance of the longitudinal pipe. The muffler is fed through the flow source on the left.

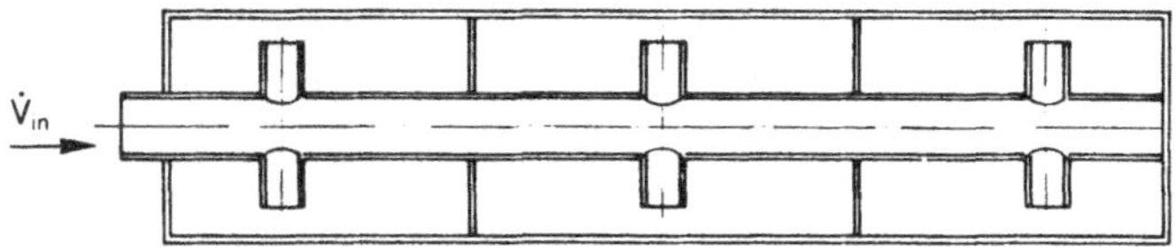

Fig. 7.1. Cross-section of a muffler.

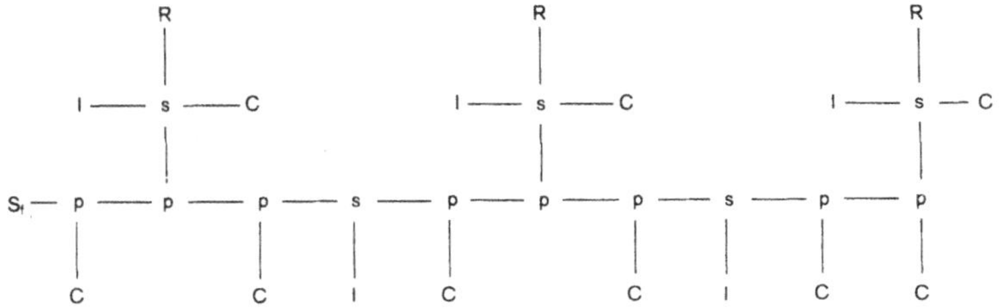

Fig. 7.2. Systematic bond graph of the muffler of Fig. 7.1.

The bond graph of Fig. 7.3 is obtained by contracting several p-junctions of Fig. 7.2, where now some C-elements represent the action of two capacitances. Furthermore the bonds are numbered consecutively from 1 to 22, and the positive power directions are selected. This allows the designation of all elements without ambiguity, since, for example C 14 is the C-element on bond 14.

The structure of the bond graph is transmitted through the punched card for the computer in the following form, requiring three punched cards:

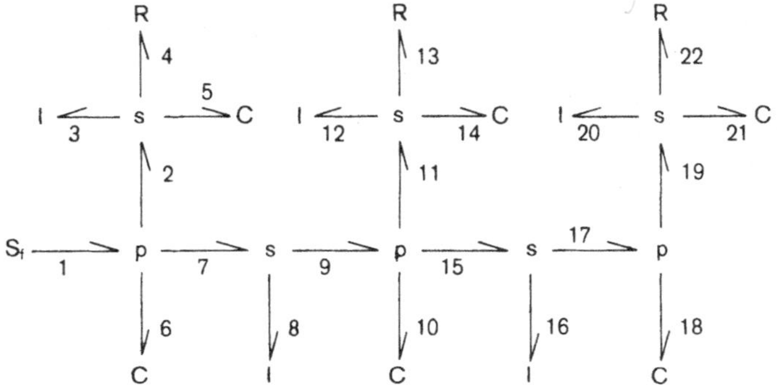

Fig. 7.3. Simplification of Fig. 7.2 with positive power direction.

SF 1, I 3, R 4, C 5, C 6, I 8, C 10, I 12, R 13, C 14, I 16,
C 18, I 20, C 21, R 22, P 1 2 6 7, S 2 3 4 5, S 7 8 9, (7-3)
P 9 10 11 15, S 11 12 13 14, S 15 16 17, P 17 18 19,
S 19 20 21 22.

The next step is to indicate the values of the parameters as follows on two punched cards:

I 3 = 1, I 12 = 1, I 20 = 1, I 8 = 2, I 16 = 2, C 5 = 1,
C 14 = 1, C 21 = 1, C 6 = 4, C 10 = 4, C 18 = 8, R 4 = 0,1, (7-4)
R 13 = 0,1, R 22 = 0,1

Subsequently the input and the desired output variables are programmed into the computer.

This is a linear problem and the computer first determines the matrices **A** and **B** of (7-1) and then the usually complex characteristic values and the stability of matrix **A**. The result is printed out. Subsequently the matrices **C** and **D** are calculated and printed out together with a plot of the time-dependent output variables. This concludes the simulation that can be easily repeated for other values of the parameters.

The above example has been demonstrated to the author in May 1973 at the University of California, in Davis, USA, by Prof. D. Karnopp and Mr. Ross

Allen. The latter also wrote the program based on Figs. 7.2 and 7.3 which was transcribed in the form of eqns. (7-3) and (7-4) on punched cards. The following simulation was carried out on a large computer containing the ENPORT program on magnetic tape in time sharing. The entire programming required about 30 min with trial runs and the computer time was 43.7 secs.

Simulation by ENPORT programs supplies the solution to engineering problems of such a complexity that they can no longer be computed by hand. However, according to Karnopp the ENPORT program requires relatively much computer storage capacity and is therefore most suitable for systems of the 5th to 50th order. With large systems of, say, 100th to 1000th order, other programs are preferable, but they require much more time for preparation.

Newer versions on ENPORT can be used to obtain directly the algebraic expressions for the transfer function. Finally, there is a direct connection to the programs written by Prof. L. Melza which allows automatic plotting of root locus curves and Bode diagrams (ref. 14).

Apart from the ENPORT program, bond graphs are exceedingly suitable to prepare simulation by different computer programs, including the so-called matching analysis of complex mechanical engineering components like heat exchangers, turbines, and compressors after F. J. Wallace.

CHAPTER 8

Selected Questions

8.1. Inertia and Resistance Fields

8.1.1. *General*

Whilst C-fields have been treated already in § 5.3 using the example of an electric capacitor with movable plates, it should be again noted here that all elements with a determined single-valued functional relation between displacements and efforts are C-fields in the sense of bond graphs. This includes deformable mechanical structures such as beams if the mass forces are neglected. Such beams have cross-couplings in general, or influences between the bonds and the Maxwell relations (5-13) and (5-14) are known for beams and elastic structures as the reciprocity theorem of Maxwell.

An inertia field, on the other hand, has a determined static relation between the momenta and the flows of the bonds where, again, cross-couplings are possible. An example is a rigid beam with inertia where forces attack at certain suspension points. Each suspension point constitutes a bond with a general relation between the momenta and flows, in this instance the velocities.

Figure 8.1 contains the example of a beam constrained to vertical movements and to small rotations around an axis perpendicular to the plane of the paper. The schematic (top) and the bond graph (bottom) have both suspension points 1 and 2 with the forces F_1 and F_2 and the velocities v_1 and v_2. The I-element on the left represents the movement of the centre of gravity with the mass m, the I-element on the right the rotation. The transformers with the ratio L/2 (-L/2 on the left) shows the lever action of the suspension

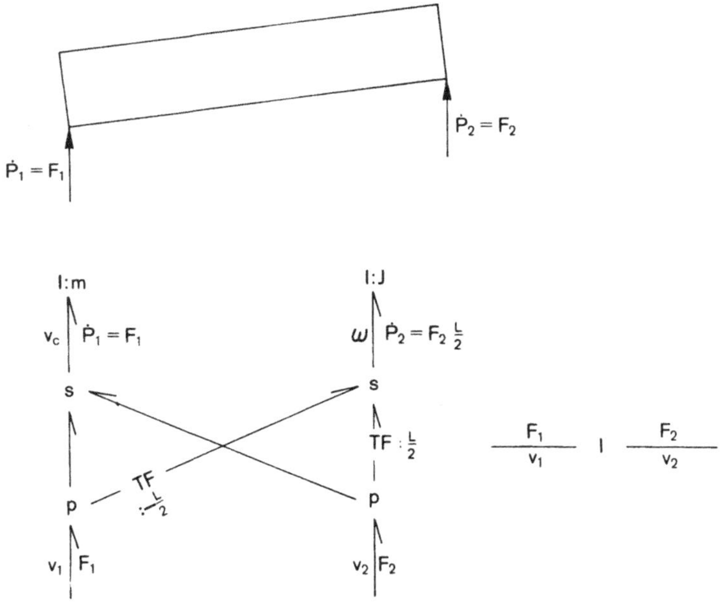

Fig. 8.1. Beam constrained to vertical movements and small rotations in the plane of the paper; schematic (top), implicit I-field (bottom left), and explicit I-field (bottom right).

forces. The force of rotation is equal to the force responsible for the movement of the centre of the mass, but rotation itself is substracted from this movement. Hence, the lower p-junctions are appropriate to combine both. The s-junctions show that the displacements of both suspension points are coupled and that forces and torques are added. Figure 8.1 also contains a short-hand representation as I-field (bottom right).

The efforts of the I-elements of Fig. 8.1 have been denoted as time derivatives $\dot{p}_1$ of the linear and $\dot{p}_2$ of angular momentum in order to emphasize that the I-elements contain a static relation between momenta and flows.

The example of the beam illustrates the following two possible forms of I-fields.

(1) As explicit I-fields, Fig. 8.1 (bottom right) as very compact representation without any detail.
(2) As implicit I-field, Fig. 8.1 (bottom left) composed of simple I-elements, junctions, and transformers.

The implicit I-field is more complex but gives worth-while insight as to how the suspension forces and velocities are transformed into the other variables like torque and rotation frequency. It contains in addition to the I-elements a junction structure that does not store energy and where the energy flux is conserved at all times. Such a junction structure is something of a switch yard (after Karnopp) that determines how the energy is distributed on the storing I-elements. Similar relations apply to C-fields. Further examples of I-fields are lossless inductors where the cross-couplings are represented by the mutual inductances.

The designation implicit and explicit fields was introduced by Karnopp. The author would prefer to interchange them since the bond graph of Fig. 8.1 (bottom left) gives more explicit information than the one at bottom right.

8.1.2. *Energy storage and reciprocity of I-fields*

The stored energy can be determined from the energy fluxes in each bond by integration with respect to time as follows:

$$E = \int (e_1 f_1 + e_2 f_2)\, dt = \int f_1 dp_1 + f_2 dp_2 \qquad (8\text{-}1)$$

where $e_1 = \dot{p}_1$ and $e_2 = \dot{p}_2$.

The substitutions $e dt = dp$ are possible since in I-fields all variables depend only on the momenta by definition. Equation (8-1) is also valid for more than two bonds.

The stored energy is consequently a natural function of the momenta and the flows are determined by derivation of the stored energy with respect to the corresponding momenta:

$$f_1 = E^{|p_1}_{-p_2} \qquad f_2 = E^{|p_2}_{-p_1} \,. \qquad (8\text{-}2)$$

A further mixed derivation gives the following result which corresponds to the Maxwell relations of C-fields:

$$f_1^{|p_2} = f_2^{|p_1} . \tag{8-3}$$

I-fields have consequently properties analogous to C-fields. In particular, the stored energy corresponds to the area below the characteristics between momentum and flow in Fig. 8.2 and is a natural function of the momentum as mentioned. The area between the characteristics and the Y-axis is the coenergy, a natural function of flow, in the case of mechanical I-elements of the velocity.

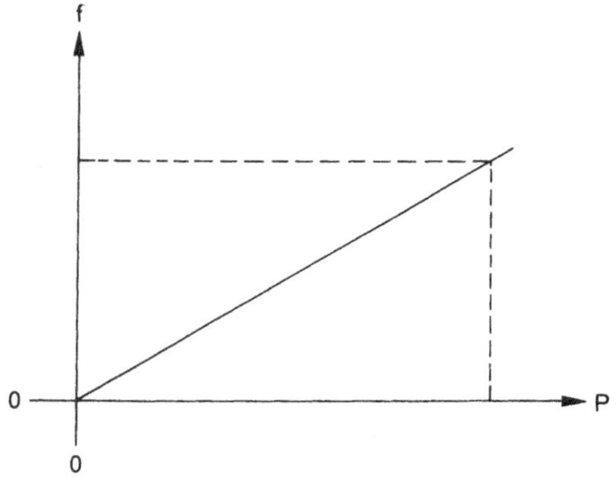

Fig. 8.2. Characteristic between flow and momentum of a linear I-field.

The important difference of energy and coenergy was seemingly introduced by ref. 10 into mechanics. Since the characteristics between momentum and velocity is linear in non-relativistic mechanics, both energies have the same numerical value.

8.1.3. *Reciprocity of R-fields*

An R-field with an arbitrary number of bonds is a multiport with static relations between efforts and flows. In particular, these relations are not influenced by the momenta and displacements (time integrals of efforts and flows). Strictly this definition includes also junction structures, junctions, transformers, and sources, but we shall consider here only R-fields that dissipate power (transform into entropy flow and absolute temperature, chapter 6).

Figure 8.3 contains an R-field with three bonds in the so-called resistance causality (left) and in the conductance causality (right).

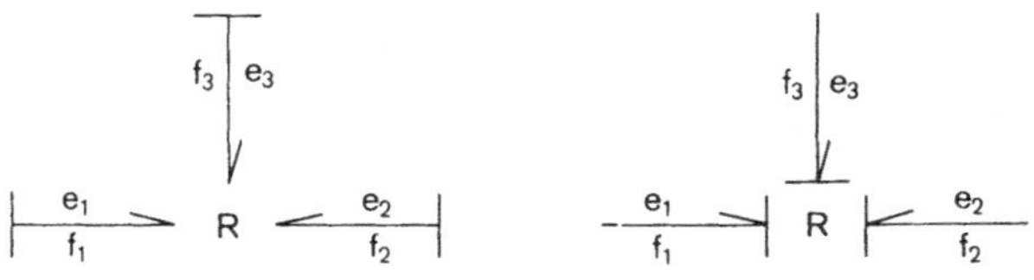

Fig. 8.3. R-fields with three bonds in different causalities.

The former can be written in the non-linear case with a set of functions F_i and the linear case with the resistance matrix z_{ij}.

$$e_i = F_i\,(f_1, f_2, f_3) \qquad e_i = \sum_j z_{ij}\, f_j \tag{8-4}$$

For the conductance causality a different set of functions G_i and the admittance matrix Y_{ij} are used:

$$f_i = G_i\,(e_1, e_2, e_3) \qquad f_i = \sum_j y_{ij}\, e_j. \tag{8-5}$$

The transfer from one to the other causality is obtained in the linear case simply by inversion of the matrices. It should also be noted that R-fields can be written in mixed causality, i.e. in resistance causlity in some bonds and in conductance causality in others.

R-fields do not have a stored energy function which is the basic reason behind the Maxwell relations of the I- and C-fields. Nevertheless, they have useful symmetry relations. In particular, the resistance and conductance matrices are symmetrical with implicit R-fields built from R-elements and junction structures without gyrators

$$z_{ij} = z_{ji} \qquad y_{ij} = y_{ji}\,. \tag{8-6}$$

The relation (8-6) is called Onsager symmetry (or reciprocity) by Karnopp in analogy to the symmetry postulated 1931 in the thermodynamics of flow processes (§ 8.3). If the R-fields contain gyrators the above symmetry generally does not apply, but then the matrices are sometimes antisymmetrical, expressed by a minus sign in (8-6) called Casimir symmetry. Casimir symmetry is found also for R-fields without gyrators in mixed causality (ref. 2, p. 8.32).

R-fields always dissipate energy, meaning that the sum of all energy fluxes is positive into the field. Nevertheless, it is quite possible that on one or the other bond, power comes out as long as a larger energy flux enters the other bonds. Including thermal effects the lost power appears (in the steady state) on a thermal bond as entropy flow and absolute temperature. Hence the R-field becomes an RS-field (§ 8.3).

8.2. Linear Two-ports in Different Causalities

Linear two-ports are used frequently in electronics where they were formerly called quadrupoles according to their four electric terminals. The relation between the flows and efforts in both bonds can be described in matrices in the most general case. Contrary to the fields of §8.1 the nature of the elements in the present two-ports is unlimited and they can contain arbitrary R-, I-, or C-elements (resistors, inductors and capacitors).

Two-ports can be represented in different forms by matrices corresponding to different causalities according to Table 8.1. The bond graph of the two-ports is denoted by TP. The first line of the table does not admit any causal representation, since both e_2 and f_2 are the input variables. They produce the variables e_1 and f_1 by multiplication with the transmission matrix, also called chain matrix. This representation is useful for several two-ports in cascade because the combined transmission matrix is obtained by multiplication of the single matrices.

The elements of the matrices in Table 8.1 are denoted by a letter that indicates their dimensions (in the sense of the theory of physical variables). In particular, the y-elements are admittances (ratios of effort to flow) and the n-elements dimensionless factors. The indices refer to the positions of the element in the matrix.

Table 8.1. *Representation of linear two-ports in different causalities by matrices*

M Transmission	$\frac{e_1}{f_1}$ TP $\frac{e_2}{f_2}$	$\begin{matrix} e_1 \\ f_1 \end{matrix} = \begin{vmatrix} n_{11} & z_{12} \\ y_{21} & n_{22} \end{vmatrix} \times \begin{vmatrix} e_2 \\ f_2 \end{vmatrix}$
Z Impedance	$\vdash\frac{e_1}{f_1}$ TP $\frac{e_2}{f_2}\dashv$	$\begin{matrix} e_1 \\ e_2 \end{matrix} = \begin{vmatrix} z_{11} & z_{12} \\ z_{21} & z_{22} \end{vmatrix} \times \begin{vmatrix} f_1 \\ f_2 \end{vmatrix}$
H Imittance	$\vdash\frac{e_1}{f_1}$ TP $\vdash\frac{e_2}{f_2}$	$\begin{matrix} e_1 \\ f_1 \end{matrix} = \begin{vmatrix} z_{11} & n_{12} \\ n_{21} & y_{22} \end{vmatrix} \times \begin{vmatrix} f_1 \\ e_2 \end{vmatrix}$
G Adpedance	$\frac{e_1}{f_1}\dashv$ TP $\frac{e_2}{f_2}\dashv$	$\begin{matrix} f_1 \\ e_2 \end{matrix} = \begin{vmatrix} y_{11} & n_{12} \\ n_{21} & z_{22} \end{vmatrix} \times \begin{vmatrix} e_1 \\ f_2 \end{vmatrix}$
Y Admittance	$\frac{e_1}{f_1}\dashv$ TP $\vdash\frac{e_2}{f_2}$	$\begin{matrix} f_1 \\ f_2 \end{matrix} = \begin{vmatrix} y_{11} & y_{12} \\ y_{21} & y_{22} \end{vmatrix} \times \begin{vmatrix} e_1 \\ e_2 \end{vmatrix}$

The second line of Table 8.1 contains the impedance representation, where both flows are input variables and the elements of the matrix are impedances. It follows the imittance representation, also called H-representation, with f_1 and e_1 as input variables. In this form the matrix elements are also called H-elements (hybrid members), since they consist of admittances, impedances, and dimensionless factors. It is used frequently in electronics for bipolar transistors with small signals. The fourth line contains the adpendance representation with mixed input variables and the last line the admittance representation of the two-ports, where all matrix members are admittances and the input variables are efforts.

The relatively unfamiliar terms imittance and adpedance are adapted from R. Rosenberg (ref. 2, chapter 10). They denote simply the action of the two-port in certain causalities, as represented by the imittance and adpedance matrices on the right of the 3rd and 4th lines of Table 8.1, and are a natural extension of the more familiar impedance and admittance matrices in the same table.

All these matrix representations of two-ports have advantages and disadvantages and are used in electronic circuit theory. Bond graphs show that they are generated by different selections of causality in universal linear two-ports.

Table 8.2 shows how the different matrices can be determined from the transmission or M-matrix with the members a, b, c, and d. The quantity Δ is the determinant of the four elements $\Delta = ad - cb$.

Table 8.2. *Calculation of different matrices from the transmission-matrix with the members a, b, c, and d.*

→ M →	$M = \begin{vmatrix} a & b \\ c & d \end{vmatrix}$
|→ Z →|	$Z = \frac{1}{c}\begin{vmatrix} a & -\Delta \\ 1 & d \end{vmatrix}$
|→ H →	$H = \frac{1}{d}\begin{vmatrix} b & \Delta \\ 1 & -c \end{vmatrix}$
→| G →|	$G = \frac{1}{a}\begin{vmatrix} c & \Delta \\ 1 & -b \end{vmatrix}$
→| Y |→	$Y = \frac{1}{b}\begin{vmatrix} d & -\Delta \\ 1 & -a \end{vmatrix}$

$\Delta = ad - cb$

Two-ports are frequently combined, especially in electrical circuits. Depending on the junctions and the causalities the matrices for the combined elements can be easily calculated from Table 8.3. Both two-ports are denoted by A and B, the ingoing bond is labelled 1 and the outgoing bond labelled 2.

As indicated on the first line of Table 8.3, the transmission matrix for a cascade (series) circuit of two-ports is obtained by matrix multiplication of the single transmission matrices. If both two-ports are connected through s-junctions the common Z-matrix results from addition of the elements of the simple Z-matrices. Similarly with two p-junctions the common Y-matrix is obtained by addition of the elements of the single Y-matrices according to the last line of Table 8.3. This procedure is a generalization of the method of addition, the impedances in series circuits, and the admittances in parallel circuits.

With mixed junctions the elements of H- and G-matrices can be added according to the third and fourth line of the table. Finally, it should be noted that very simple elements such as junctions with a one-port on one bond can be described by a matrix between the two remaining bonds.

Table 8.3. *Different circuits of linear two-port and corresponding connections through matrices*

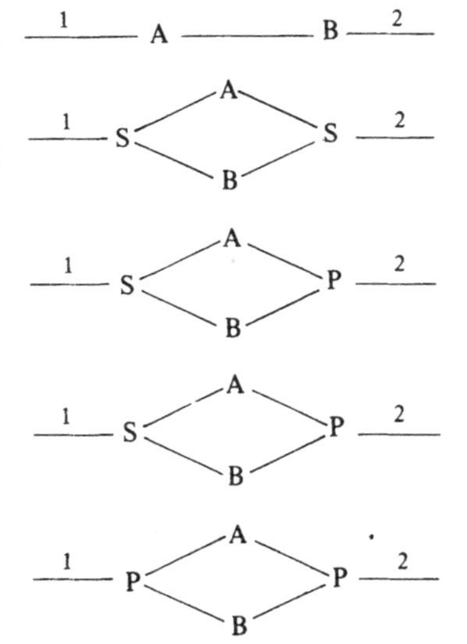

M	$M_{12} = A \times B$
Z	$Z_{12} = Z_A + Z_B$
H	$H_{12} = H_A + H_B$
G	$G_{12} = G_A + G_B$
Y	$Y_{12} = Y_A + Y_B$

From the above the impression emerges that bond graphs do not bring anything fundamentally new for the highly developed electronic circuit theory, but they allow the construction of the usual forms of the matrices of linear two-ports systematically by different selections of causalities.

The relations of this section between bond graphs and matrices shows the similarity of both methods. Karnopp in particular emphasizes the fact that bond graphs correspond to the method of "tearing of matrices of Gabriel Kron" (ref. 18).

8.3. Thermodynamics of Flow Processes

The thermodynamics of flow processes (irreversible thermodynamics) is frequently considered as difficult, but its fundamentals can be represented clearly by bond graphs.

As mentioned in Chapter 6, conventional thermodynamics – also called thermostatics – refers to matter and working substances behaving like C-fields with a cross-coupling between thermal and other bonds. Any resistance or functional relation between effort and flow is carefully avoided, in particular friction and leakage of pistons and entropy flow under finite temperature drop.

Thermodynamics of flow processes deal specifically with resistances or R-elements and is thus a natural extension of the entropy generation of § 6.3 and especially of the RS-fields of Figs. 6.6 and 6.7. One important problem arises with two or more constant flows through a substance, e.g. an entropy flow and an electric current through a bar. We limit ourselves to the steady state and exclude storage effects (C-elements). Figure 8.4 contains a schematic (left) and a bond graph (right) with RS-fields where power enters through both vertical bonds. Just as with heat conduction (Fig. 6.7), the dissipated power leaves as newly generated entropy

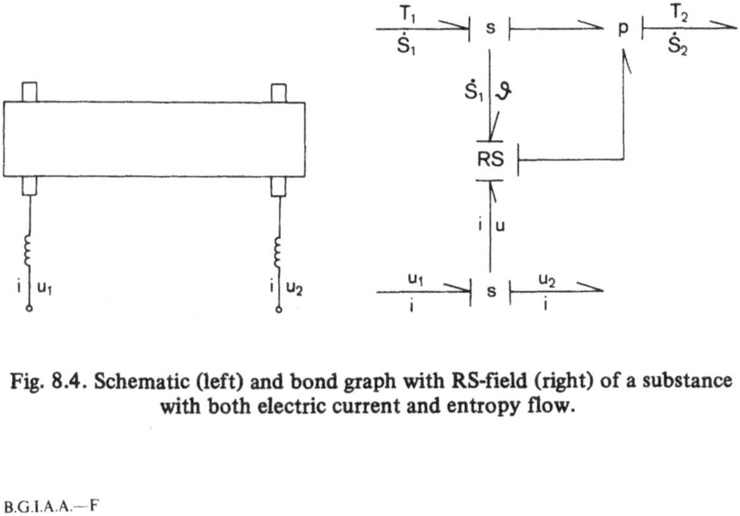

Fig. 8.4. Schematic (left) and bond graph with RS-field (right) of a substance with both electric current and entropy flow.

flow through the central horizontal bond and enters the bond 2 on the right through the p-junction. An important quantity is the power dissipated for entropy generation in analogy to the stored energy in C-fields:

$$\dot{E} = \vartheta \dot{S}_1 + ui = e_1 f_1 + e_2 f_2 . \tag{8-7}$$

Here $\vartheta = T_1 - T_2$, $u = u_1 - u_2$, are the temperature and voltage drop. The second, general form of (8-7) with the efforts e_1 and e_2 and the flows f_1 and f_2 has been included for completeness.

Figure 8.4 contains an RS-field similar to § 8.1 where one flow is the entropy flow. As in a network of electrical resistors, the bonds are cross-coupled in many substances in that, for example, f_1 depends not only on e_1 but also on e_2. In particular, the entropy flow depends in many substances not only on the temperature difference, but also on the applied voltage. The analysis is usually limited to the linear case (small voltages and temperature differences) where the efforts vanish if the flows are zero. Then the flows can be written as function of the efforts and the efforts as function of the flows as follows:

$$\begin{aligned} \dot{S} &= y_{11}\vartheta + y_{12}u; \qquad & \vartheta &= z_{11}\dot{S} + z_{12}i, \\ i &= y_{21}\vartheta + y_{22}u; \qquad & u &= z_{21}\dot{S} + z_{22}i. \end{aligned} \tag{8-8}$$

The first form of (8-8) corresponds to the causality of Fig. 8.4 and the second form has a causality with flows as input variables.

According to (8-8), the vector of the flows with the components $\dot{S}$, i is obtained by multiplication of the vector of efforts with the components ϑ, u with the matrix **Y**. The matrix **Z** for determination of the efforts from the flows is the inverse of the **Y**-matrix. Using matrices is especially useful with three or more coupled flows.

The coefficients z_{ij} can be calculated by interpreting (8-8) (right) as linear system for ϑ and u and to solve it by determinants (the Cramer rule):

$$z_{11} = \frac{y_{22}}{|y|} \qquad z_{12} = -\frac{y_{12}}{|y|} \qquad z_{21} = -\frac{y_{21}}{|y|} \qquad z_{22} = \frac{y_{11}}{|y|} \tag{8-9}$$

where $|y| = y_{11}y_{22} - y_{12}y_{21}$ is the determinant of **Y**.

In most flow processes there holds the following symmetry relation just as in electric R-fields:

$$y_{12} = y_{21} \qquad z_{12} = z_{21}$$

The above relation is called the Onsager symmetry condition and plays an important role in thermodynamics of flow processes. Originally it was considered "mysterious", later derived from statistical mechanics, but we shall derive it below from the 2nd law of thermodynamics.

The matrices **Z** and **Y** are then symmetric. The off-diagonal elements are frequently different from zero and give the coupling of the electrical and thermal bonds. An electric current then entrains an entropy flow even without temperature drop, and the ratio of these flows is denoted by

$$\frac{\dot{S}}{i} = \frac{y_{12}}{y_{22}} = \epsilon. \tag{8-10}$$

Further easily observable parameters on the bar with entropy flow and electric current are the electrical admittance with zero temperature drop and the entropy conductance σ with vanishing current.

$$\gamma = \left(\frac{i}{u}\right)_{\vartheta = o} = y_{22}, \qquad \sigma = \left(\frac{\dot{S}}{\vartheta}\right)_{i = o} = y_{11} - y_{12}\frac{y_{21}}{y_{22}}. \tag{8-11}$$

These three observable parameters are sufficient to determine together with the Onsager conditions the four coefficients of the matrices Y and Z.

The power conversion and the efficiency of the bar represented by the RS-field of Fig. 8.4 is interesting. The power fluxes entering are

$$\dot{E}_{el} = ui = u(y_{21}\vartheta + y_{22}u), \qquad \dot{E}_{th} = \vartheta(y_{11}\vartheta + y_{12}u). \tag{8-12}$$

Next the voltage is expressed in terms of ϑ and i by (8-8) bottom left and the resulting y coefficients are eliminated by (8-10) and (8-11) without imposing the Onsager condition yet. The result is

$$\dot{E}_{el} = \frac{i^2}{\gamma} - \frac{y_{21}}{\gamma}\,\vartheta, \quad \dot{E}_{th} = \sigma\vartheta + \epsilon i\vartheta . \tag{8-13}$$

Note that for electrical power output the power inflow E_{el} must become negative (also the voltage in (8-12)). The thermoelectric conversion efficiency is

$$\eta_e = -\frac{\dot{E}_{el}}{\dot{E}_{th}} = \frac{y_{21}\, i\vartheta/\gamma - i^2/\gamma}{\epsilon i\vartheta + \sigma\vartheta^2} . \tag{8-14}$$

The device of Fig. 8.4 can also be run as an entropy pump, delivering entropy at higher temperature by absorbing electrical power. For this the temperature difference must reverse its sign, and the electrical power is always positive by (8-13) with $\vartheta < 0$. Introducing $\vartheta' = -\vartheta$ ($\vartheta' > 0$), the electrothermal efficiency becomes

$$\eta_{th} = -\frac{\dot{E}_{th}}{\dot{E}_{el}} = \frac{\epsilon i \vartheta' - \sigma\vartheta'^2}{y_{21}\, i\vartheta'/\gamma + i^2/\gamma} . \tag{8-15}$$

With small loss terms $\sigma\,\vartheta^2$ and i^2/γ (small entropy admittance and high electrical conductance), one of the conversion efficiencies would be greater than 1 unless the following condition is fulfilled (the Onsager symmetry condition):

$$\frac{y_{21}}{\gamma} = \epsilon, \quad y_{12} = y_{21} . \tag{8-16}$$

The conversion efficiencies can be at most equal to 1 by the second law of thermodynamics otherwise an entropy flux would enter the source section of the RS-field and be destroyed. Hence the Onsager conditions follow directly from the second law and then there is no minimum value for the loss terms. They represent simply the Joule effect (i^2/γ) and the thermal conduction loss power ($\sigma\vartheta^2$).

With the Onsager symmetry the thermoelectric conversion efficiency simplifies to

$$\eta_e = \frac{\epsilon i\vartheta - i^2/\gamma}{\epsilon i\vartheta + \sigma\vartheta^2} = \frac{1 - i/\vartheta\epsilon\gamma}{1 + \sigma\vartheta/i\epsilon} . \tag{8-17}$$

The material properties $1/\epsilon\gamma$ and σ/ϵ should be small for high efficiency. With a given material it is possible to choose the operating ratio i/ϑ such that the efficiency has a maximum. Its value depends only on the dimensionless product $\sigma/\epsilon^2\gamma$.

In order to obtain the total efficiency, the Carnot factor has to be added to (8-17):

$$\eta_{tot} = -\frac{\dot{E}_{el}}{T_1\dot{S}} = \frac{\epsilon i\vartheta - i^2/\gamma}{\epsilon i\vartheta + \sigma\vartheta^2}\,\frac{T_1 - T_2}{\Gamma_1}. \qquad (8\text{-}18)$$

The loss power appears as new entropy flow and is injected by the source part of the RS-field

$$T_2\Delta\dot{s} = \sigma\vartheta^2 + i^2/\gamma. \qquad (8\text{-}19)$$

The device with coupled electric currents and entropy flows is used frequently, both for thermoelectric power generation and as an entropy pump for the cooling of domestic refrigerators and electronic instruments. It is responsible for several known efforts in physics, e.g. the Peltier effect.

In metals, where both electric charge and entropy are transported by electrons, electric and thermal (entropy) conductance are proportional and largely independent of chemical composition and temperature (Wiedemann-Franz, rule ref. 9).

The thermoelement is another application, where two different substances are connected in series according to Fig. 8.5. If the interfaces are kept at different temperatures T_h and T_n, electric current and voltage are experienced and can be measured by a voltmeter.

The injection of the newly generated entropy is in reality distributed evenly in the substance. Consequently the finite RS-field is only an approximation but sufficient for the explanation of principles. For more accuracy it would be necessary to connect many small RS-fields and add or integrate their effects (ref. 9).

Similar coupled flow processes are found between water and electric current in membranes, in the so-called electro-osmosis, with chemical reactions and as thermodiffusion. They can all be described by RS-fields.†

† Although the symmetry was derived here for the linear.

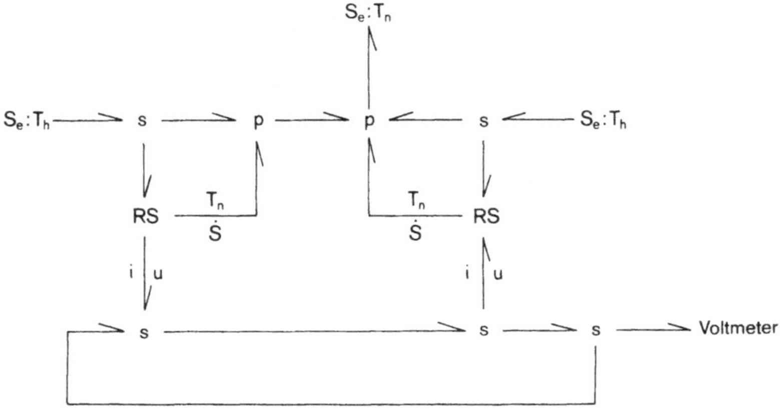

Fig. 8.5. Series circuit of two substances of Fig. 8.4 in a thermoelement.

8.4. Electromechanical Components

This section is devoted to the more detailed description of components linking the electric, magnetic, and mechanical energy domains by bond graphs, whilst their application as motors and signal transducers has already been discussed in §§5.1 and 5.3.

The starting point is the magnetic field with the magnetic tension λ (also called magnetomotive force or mmf) corresponding to the ampere turns of a field coil and the magnetic flux Φ (really a displacement, see below). It should be noted that the magnetic tension is basically equal to the magnetic field H multiplied (or integrated over) the length of the magnetic circuit. The magnetic flux is equal to the flux density B multiplied with the area. The energy increase of a magnetic field is given by the magnetic tension and the increase of magnetic displacement which justifies this classification.

According to the laws of magnetism there is a functional relation between magnetic tension and displacement (i.e. between H and B) which is represented by a C-element of bond graphs. With constant magnetic permeability this C-element is linear, otherwise non-linear, especially with saturation. This magnetic circuit is a C-field only if hysteresis is negligible. If necessary it can be taken into account by a frequency dependent R-element, but the following considerations are limited to pure C-fields or magnetic circuits without hysteresis.

According to the induction law the electric voltage is proportional to the time derivative of the magnetic flux and the magnetic tension to the electric current

$$u = n \frac{d\Phi}{dt} \qquad \lambda = n\,i \tag{8-20}$$

where n is the number of turns.

It would be necessary in principle to introduce the so-called Fleischmann constant γ in a way that instead of the number of turns the ratio n/γ appears. This is needed to express the relationship between magnetic and electric energy correctly in the sense of the theory of physical dimensions. However, we have set here as usual $\gamma = 1$ (ref. 21).

The induction law corresponds to a gyrator due to the connection of electric tension and magnetic flux derivative and of magnetic tension and electric current by (8-20). The magnetic circuit appears in Fig. 8.6 (left) as schematic with windings on an iron core and (right) as gyrator with following C-element for the relation between magnetic tension and flux. In the linear case, i.e. with constant permeability, the value of the C-element is $C = \mu A/l$:

$$u = nCn \frac{di}{dt} = \frac{A\mu}{l} n^2 \frac{di}{dt} = L \frac{di}{dt} \tag{8-21}$$

where A is the cross section, l = the length of the magnetic circuit, and $L = A\mu n^2/l$ the well-known inductance.

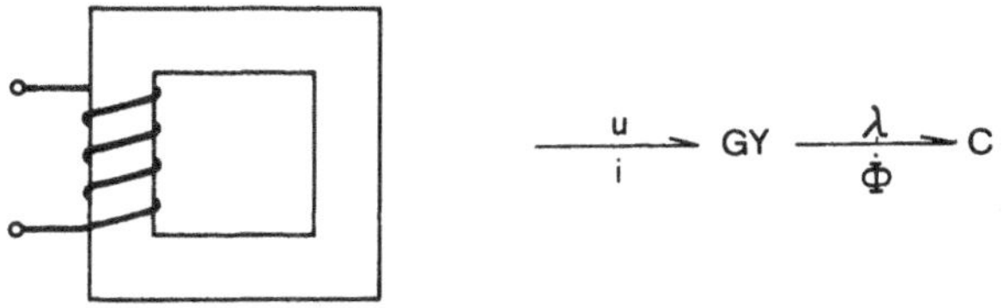

Fig. 8.6. Magnetic circuit; schematic (left) and bond graph with gyrator and C-element (right).

Magnetic circuits with movable iron members or magnetic actuators are a C-field with a gyrator on the electrical side. As soon as the relation between magnetic tension and flux is influenced by the position of the iron core, we have cross-coupling between the bonds and obtain forces as described in § 5.3. Figure 8.7 shows a schematic (left) and the bond graph (right) of a magnetic actuator with a C-field. By this figure even these energy convertors are explained or at least described by the standard elements of bond graphs.

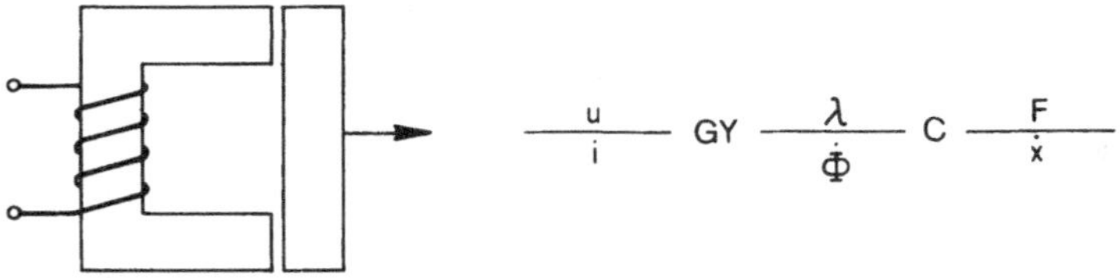

Fig. 8.7. Magnetic circuit with movable iron; schematic (left) and bond graph with gyrator and C-field (right).

The bond graph of Fig. 8.7 is a detailed explanation of the IC-field of Fig. 5.14, which describes the action of the electromagnetic signal transducer more compactly. The IC-field is an explicit field in the sense of Karnopp and Fig. 8.7 an implicit field. It is more complex but gives considerable insight into the physics of the device.

The quantity Φ conventionally called magnetic flux, is really a displacement and not a flow because otherwise the parts of a magnetic circuit would be R-elements (magnetic resistors). This would be mathematically possible but physically incorrect since R-elements always dissipate power. A static relation between magnetic tension and magnetic flux (displacement) does not dissipate any power corresponding to the properties of a C-field. In particular a constant magnetic field can be generated by permanent magnets without power, and an electromagnet does not dissipate any power in the steady state if the resistance of the windings is negligible, e.g. with superconductive electromagnets.

Figure 8.8 shows a magnetic circuit with two windings, coupled magnetically but having a certain stray flux, as in an electric transformer. The bond graph is set up systematically by associating to each branching point of the flows (point of constant magnetic tension) p-junctions and by applying the C-elements and the gyrators of both windings in between through s-junctions. The C-element C_3 gives the stray flux due to the magnetic tension differences between both p-junctions.

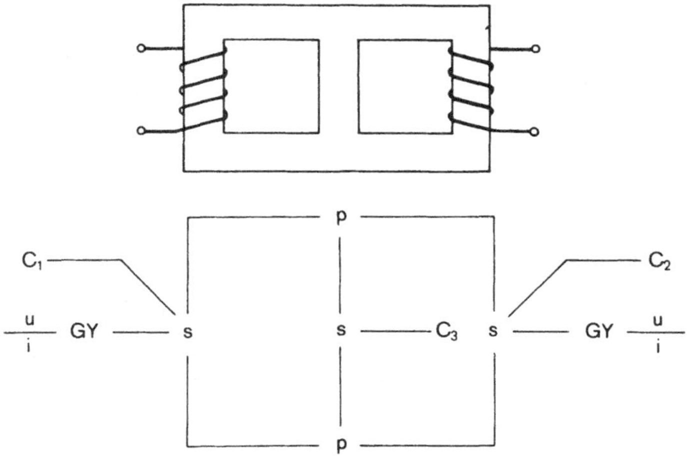

Fig. 8.8. Electrical transformer with stray flux.

Figure 8.9 is obtained by setting the magnetic tension of the lower p-junction of Fig. 8.8 to zero. It can be considered as an extension of Fig. 8.6 and gives complete information about currents

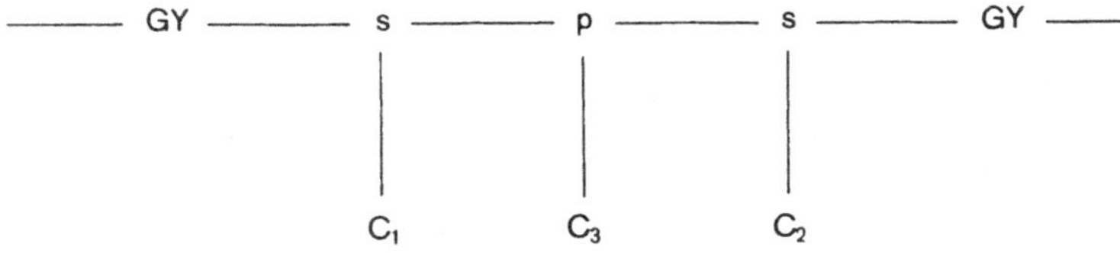

Fig. 8.9. Simplified bond graph of the electrical transformer with vanishing magnetic tension in the lower p-junction of Fig. 8.8.

and voltages in both windings, i.e. the self and mutual inductances. Losses and hysteresis could be represented by additional R-elements. Due to the reciprocity of C-fields, both mutual inductances are equal.

Figure 8.10 shows a bond graph of a d.c. electric motor with shunt excitation, obtained from Figure 5.8 by replacing the magnetic circuit with a gyrator and a magnetic C-element on the bottom. It shows a little more in detail the operation of the motor with the magnetic displacement as conversion parameter of the modulated gyrator (top right). It carries out the essential conversion of electric into mechanical power.

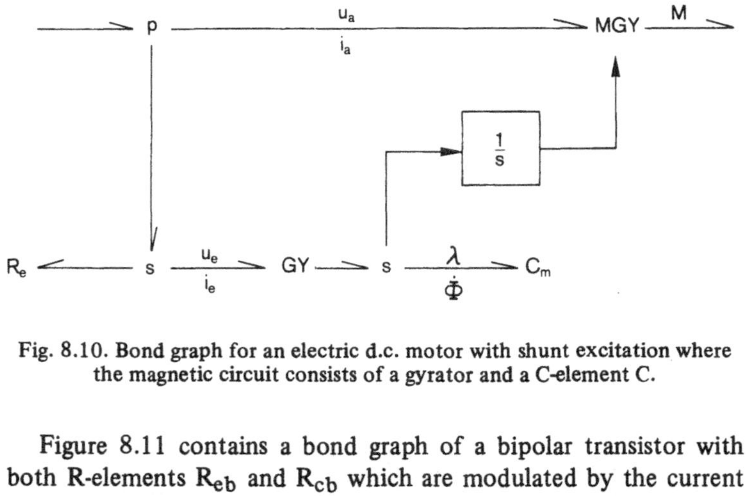

Fig. 8.10. Bond graph for an electric d.c. motor with shunt excitation where the magnetic circuit consists of a gyrator and a C-element C.

Figure 8.11 contains a bond graph of a bipolar transistor with both R-elements R_{eb} and R_{cb} which are modulated by the current in the other bonds. This is in reality the effect of the transition of the minority carriers through the basis zone. The bond graph does not give detailed information but it shows clearly the fundamental operation.

The bond graphs of this section allow representation of operation of different electrical components and machines in an intuitively appealing manner. They show in particular how the different equations are related to each other. The fact that the equations of a large part of engineering science have a very similar form is the basic reason that the few standard elements of bond graphs have such a large range of application.

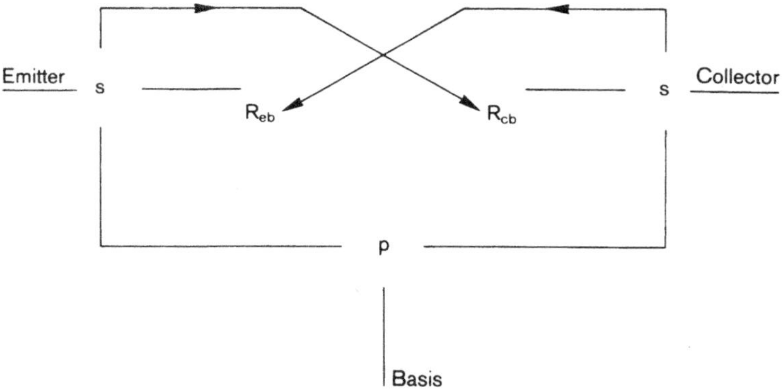

Fig. 8.11. Bond graph of a bipolar electronic transistor.

8.5. Systems with Distributed Parameters

Until now bond graphs were used only to represent systems built from a finite number of elements, each describing an action at a certain place. Such systems are called systems with concentrated or lumped elements and correspond mathematically to ordinary differential equations.

Many systems have continuously distributed elements and are described mathematically by partial differential equations. This is always the case if the length of electric or mechanical waves approaches the size of a real component. The treatment of such linear systems is a classical problem of mathematical physics and uses the so-called normal modes (also called characteristic functions or eigenfunctions) selected according to the boundary conditions. The complete solution is a superposition of many normal modes. Each of these modes correspond to a resonance circuit and incorporates C- and I-elements.

We shall use as an example a hydrostatic transmission with a (rubber) hose between pump and motor as shown in Fig. 8.12, similar to Fig. 4.18, where the hose was represented by the central

C-element. Due to the soft walls, the velocity of sound decreases to about 300 m/sec (ref. 5), and the wavelength is at higher signal frequencies easily comparable to the hose length. The following analysis is based on ref. 20 and uses one-dimensional longitudinal modes only which is realistic if the hose diameter is much smaller than the wavelength. Otherwise the much more complex three-dimensional modes should be used.

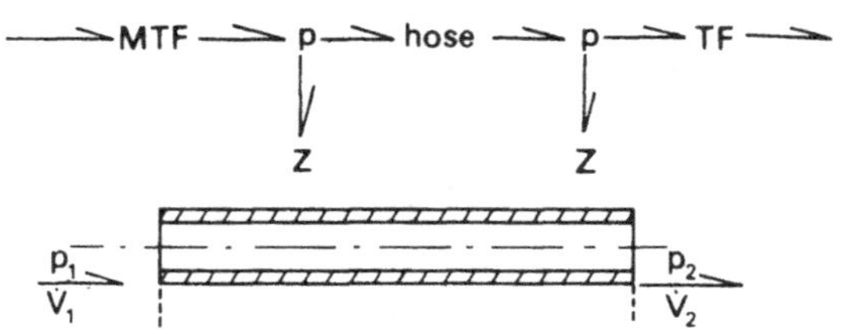

Fig. 8.12. Hydrostatic transmission with a hose between pump and motor; bond graph with a word (top) and a schematic of the hose (bottom).

The equations of motion in the pipe without friction are in the so-called acoustical approximation (small change of mass density ρ as always in hydraulic oil).

$$\frac{\rho}{A}\dot{V}|^{t} = -p|^{x} + \frac{f}{A}, \qquad \frac{p|^{t}}{\rho c^2} = -\frac{\dot{V}|^{x}}{A} + \frac{q}{A}, \tag{8-22}$$

where $c = \sqrt{(B/\rho)}$, the velocity of sound; A the cross-section of the pipe; f the external force (per length), equals external pressure times cross-section divided by length; and q the injected flow (per length).

This first line of (8-22) essentially gives the flow acceleration due to pressure gradient and external force, the second the pressure rise with compression due to flow gradient at injected flow. With suitable second derivations for time t or x, pressure or flow, can be eliminated:

$$\left.\begin{aligned} \dot{V}^{|\,|t} &= c^2 V^{|\,|x} - c^2 q^{x} + \frac{f^{|t}}{\rho}, \\ \frac{p^{|\,|t}}{c^2} &= p^{|\,|x} + \frac{\rho}{A} q^{|t} - \frac{f^{|x}}{A}. \end{aligned}\right\} \tag{8-23}$$

The upper equation is suitable for determining the flow from an external force, e.g. a pressure acting on the pipe, and works in the flow from pressure causality. The lower equation determines the pressure from an injected flow in the pressure from flow causality. This we shall do representing the injected flow by Dirac delta functions and setting f = 0:

$$\begin{aligned} q(x) &= \frac{\dot{V}_1}{L}\,\delta\left(\frac{x-x_1}{L}\right) - \frac{\dot{V}_2}{L}\,\delta\left(\frac{x-x_2}{L}\right) \\ &= \frac{\dot{V}_1}{L}\,\delta\left(\frac{x}{L}\right) - \frac{\dot{V}_2}{L}\,\delta\left(\frac{x}{L}-1\right) \end{aligned} \tag{8.24}$$

where the factors L have been added for dimensional reasons.

The first form of (8-24) is valid for flow injected at any x_1 at x_2, the second form for injection at $x_1 = o$ and $x_2 = L$, i.e. at the pipe ends.

With the injected flow, the flow $\dot{V}$ vanishes at the pipe ends, giving the boundary condition $p^{|t}(x = 0) = p^{|t}(x = L) = 0$.

The modes are the product of a time-dependent and a space-dependent factor, and the pressure results by adding the modes as follows:

$$p\,(x, t) = \Sigma\ \eta_i(t)\ \cos\frac{i\pi x}{L}, \tag{8-25}$$

where $\eta_i(t)$ is the time-dependent factor having the dimension of a pressure. Here $\cos i\pi x/L$ is the non-dimensional space-dependent factor fulfilling the boundary conditions.

An equation for η_i is obtained by inserting (8-24) and (8-25) into (8-23), the result multiplied by the factor $\cos j\pi x/L$ and integrated with respect to x from 0 to L. Due to the orthogonality of the cosinus functions, all terms except the one with j = i vanish, yielding the following equations for η after one more integration to t:

$$\left.\begin{aligned} \frac{AL}{\rho c^2}\,\eta_o &= \dot{V}_1 - \dot{V}_2\,, \\ \frac{AL}{2\rho c^2}\,\dot{\eta}_i &= \frac{i^2\pi^2 A}{2L\rho}\int \eta_i\,dt + \dot{V}_1 - \dot{V}_2\,(-1)^i\,,\,(i = 1, 2, \ldots, \infty). \end{aligned}\right\} \quad (8\text{-}26)$$

Equation (8-26) describes the building of the components η_i of the pressure from the injected flow. The first part is the zero mode and corresponds to the same C-element as on Fig. 4.18 with $B = \rho c^2$. The higher modes are the equations of I- and C-elements on a p-junction (added flows, equal pressure) with the injected flows except for the factor $(-1)^i$ that can be represented by a transformer:

$$\left.\begin{aligned} C\,\dot{\eta}_i &= \frac{1}{I_i}\int \eta_i\,dt + \dot{V}_1 - \dot{V}_2\,(-1)^i, \\ C &= \frac{AL}{2\rho c^2} \qquad I_i = \frac{2L\rho}{i^2\pi^2 A}\,. \end{aligned}\right\} \quad (8\text{-}27)$$

With higher modes, the resonance frequencies $\omega_i^2 = 1/CI$ will become higher (due to the i in the denominator of I). Therefore, for representing the pipe up to a certain frequency only a few modes will be necessary, whilst the zero mode alone is valid for very low frequencies. As already stated, it is identical to the lumped parameter represented with a single C-element. Figure 8.13 displays (8-27) as a bond graph with only the modes of order 0, 1, and 2. The transformer takes care of the minus sign in the bracket. Although (8-27) were set up using a certain causality, no causality is attached to Fig. 8.13 and this bond graph is valid in any causality.

As an illustration from hydrostatic transmission, we have for the C-element of the zero order mode, where ρ = mass density and B = the compressibility modulus:

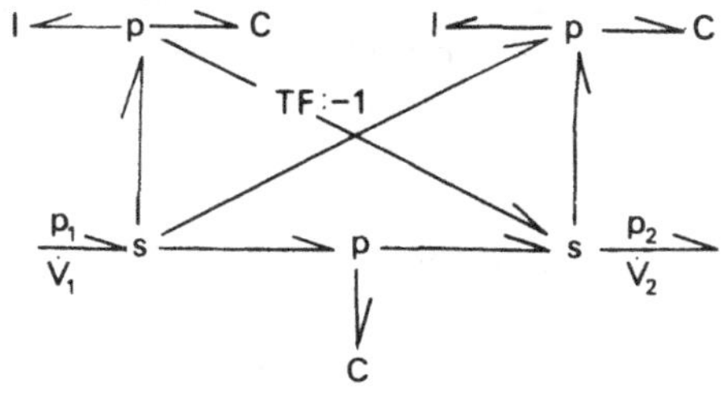

Fig. 8.13. Bond graph for the hose with the modes O (as basic C-element), 1 and 2 as resonant combination of I- and C-elements.

$$C_0 = \frac{AL}{\rho c^2} = \frac{AL}{B} = \frac{V}{B},$$

where the velocity of sound $c^2 = B/\rho$. Numerically one obtains with $B = 1.5 \times 10^9$ N/m^2, $\rho = 850$ kg/m^3 (usual values for hydraulic oil) a velocity of sound with rigid walls (steel pipes) of C = 1330 m/sec, much higher than in air.

The resonance (angular) frequency of the single modes are

$$\omega_i^2 = i^2 \pi^2 \frac{B}{\rho L^2} = i^2 \pi^2 \frac{c^2}{L^2}$$

and the frequencies and wavelength $\lambda_i = C/f_i$

$$f_i = \frac{ic}{2L}, \quad \lambda_i = \frac{2L}{i}.$$

The lowest mode (i = 1) corresponds to a wavelength of twice the pipe length and the higher modes give correspondingly shorter wavelength. In a hydrostatic drive with a pipe length of 5 m, the lowest mode resonance frequency becomes f_1 = 1330 Hz, the next twice as much, usually well above other resonance frequencies of the system.

However, with a rubber hose the velocity of sound is much smaller due to the expansion of the walls under pressure (ref. 5, § 7.2). In a typical high pressure hose this becomes c = 325 m/sec and the lowest resonance frequency f_1 = 32.5 Hz. For such systems a bond graph simulation with two modes is eminently suitable. Let us repeat that the resonance frequencies of the system are in general somewhat different due to the mode interaction.

Figure 8.13 is then the bond graph extending the simple representation of the pipe by the parallel C-element to higher signal frequencies by including the first and second-order modes. In principle as many modes as desired can be added. It should be noted, however, that the resonance frequencies of the bond graph will in general differ from the resonance frequencies of the single modes due to their coupling. In this way bond graphs display impressively the solution of partial differential equations of mathematical physics.

As a generalization, the flows can be injected not at the pipe ends but at arbitrary positions x_1 and x_2 according to the first form of (8-24). The solution then becomes

$$C\dot{\eta}_i = \frac{1}{I_i}\int \eta_i \, dt + \dot{V}_1 \cos\frac{i\pi x_1}{L} - \dot{V}_2 \cos\frac{i\pi x_2}{L}, \tag{8-28}$$

with C, I_i as in (8-27) which is a simple case of (8-28) (with $x_1 = 0$, $x_2 = L$). This is sometimes a reasonable model especially with mufflers for internal combustion engines (ref. 20).

The bond graph displaying (8-28) is similar to Fig. 8.13, only that transformers appear in the vertical and in both oblique bonds. They have the transforming ratios

$$\cos\frac{\pi x_1}{L}, \quad \cos\frac{2\pi x_1}{L}, \quad \cos\frac{\pi x_2}{L}, \quad \cos\frac{2\pi x_2}{L}.$$

Transformers always link both flows and pressures. The linking of flows of the transformers just discussed is contained in the last two terms of (8-28). The pressure linking takes place in (8-25) between the mode factor η_i and the entire pressure modes.

Alternatively it would be possible to solve the upper equation of (8-23) for the flow waves assuming injected pressures (flow from pressure causality). This yields a similar bond graph where only I- and C-elements and p- and s-junctions are interchanged, which can again be used for any external causality (ref. 20).

The solution of distributed parameter problems with normal modes can be summarized as follows:

(1) To each normal mode corresponds a resonance circuit with a certain resonance frequency coupled through a transformer to the system input bonds.
(2) The transforming ratios are the space dependent factors of each mode.
(3) For representing the system in a definite signal frequency range, only some, usually very few normal modes are needed. They can be displaced as bond graphs.
(4) The zero order mode is the lumped element solution.
(5) The bonds connecting to other parts of the system admit arbitrary causalities.

The 5th property is very helpful for simulation of, e.g. railway wagons, where the body is a subsystem with distributed elements, but the suspension (wheels, springs, and bogies) consists of lumped elements.

At the time of this writing it appears that great progress can be achieved by developing normal mode and bond-graph methods for partial differential equations. One important problem is the introduction of damping forces (R-elements) and to overcome or to live with the consequent coupling of normal modes.

8.6. Force and Velocity as Effort or Flow

In this book force and torque are taken as efforts and velocity and angular frequency as flows. This choice can be called the force–effort analogy or the direct analogy.

As mentioned in § 2.1, there also exists the dual analogy, where force and torque are flows and velocity and angular frequency are efforts. This can be called the force–flow analogy. Fortunately there is no ambiguity about the interpretation of the other variables as efforts and flows, which is always as given in § 2.1.

Mathematically it is of no consequence whether force is taken as effort or flow, but most original bond graph publications use the force–effort or direct analogy. Since some authors prefer the dual analogy, we shall compare them here including some of the consequences.

Conceptually, the direct analogy is preferred, since effort implies a force whilst flow is associated with movement as in velocity or angular frequency. For the physical interpretation it should be noted (ref. 9, p. 158) that flows reverse their direction and efforts remain unchanged if the direction of time is reversed in a thought experiment. Such time reversal could be simulated by filming a real process and projecting the film backwards.

The case for the force–flow analogy comes from systems theory, where the variables are classified as through variables and across variables (also called pervariables and transvariables). Each bond contains always one through and one across variable, and their product is the power or energy flux. This classification refers to the process of measuring the variables. An across variable like the voltage drop in a component can be measured by applying an instrument between two points. A through variable, like an electric current, requires the cutting of a connection to insert the instrument (ammeter). In this sense a velocity is an across variable measured by an instrument touching at two points, whilst a force or torque is a through variable requiring the cutting of the link (or shaft) to insert a force (or torque) meter. Consequently in the dual analogy all efforts including velocity and rotation frequency are across variables, and all flows including force and torque are through variables, which is a worth-while simplification.

The classification of across and through variables is the basis of another graphical aid of systems theory, the network graph (or linear graph) previously mentioned in § 1.4. These graphs are widely used but are much more abstract and less intuitively appealing than bond graphs. Especially for interdisciplinary systems and they also have some essential weaknesses. For an excellent introduction and comparison of both methods the reader is recommended to consult ref. 22.

The essential difference between through and across variables appears to this author not as the measuring process, but the question of the origin and the absolute values of the variables:

(1) For an across variable not the absolute value but only the value relative to some reference or datum value is significant.

As an example, an electric component is stressed only by the potential (voltage) difference and a shaft bearing by the difference of the rotation frequencies of shaft and casing.

(2) With a through variable the absolute value of the variable is important, as for example, the current in the wire and the torque in a shaft must respect absolute limits in order to avoid damage.

An exception to the above is the thermal energy, where the absolute temperature is significant in spite of the fact that temperature is an across variable.

Paynter's tetragon (§ 2.5.2) remains valid in the dual (force–flow) analogy, but it is rotated by one-half turn since effort and flow, on the one hand, and displacement and momentum and displacement on the other, also exchange places. Generally, the entire bond graph terminology is adapted to the direct analogy. If the force–flow analogy is to be used, words like momentum and displacement are best avoided and replaced by effort and flow time-integral. In this case a contact force between two bodies can be thought of as a flow of (conventional) momentum.

In order to use the force–flow analogy, the following changes have to be made in the bond graphs of this book:

(1) The p-junctions become s-junctions and vice versa.
(2) Hydraulic pumps or cylinders (Chapter 4) turn into gyrators.
(3) D.C. electric motors become transformers and the IC-field of the magnetic actuator of § 5.3.3 becomes an I-field.
(4) Mechanical C-elements and I-elements are interchanged. In particular springs become I-elements and flywheels C-elements.
(5) Mechanical C-fields and I-fields are also interchanged whilst mixed mechanical and other energy domain fields turn into IC-fields. As an example, the electrical capacitor with movable plate becomes an IC-field with I-character at the mechanical and C-character at the electric bond.

In spite of the advantages of the dual analogy for systems engineering, the direct, force–effort analogy appears to be physically more correct and should be used when the physical principles of operation of a device or, indeed, effects of physics are described by bond graphs.

Additional light is thrown on both analogies by considering energy relations of bond-graph elements. In fact they can be classified by the way they handle energy and by the number of their ports. Combining both classifications leads to Table 8.4, which is appropriately called a *portality table.* *

Table 8.4. *Portility table: bond-graph elements according to their energy properties (vertical axis) and their number of ports (horizontal axis)*

	Number of ports		
Energy handling	1	2	3 and more
Conserving		TF transformer GY gyrator [RS-field]	s-junction p-junction [RS-field]
Generating Removing	S_f source S_e source		
Dissipating	R-element	R-field	R-field
Storing	I-element C-element	I-field C-field IC-field	I-field C-field

Table 8.4 shows the number of ports, namely one, two, and three or more ports, and the different ways of handling energy. Bond-graph elements either conserve energy as transformers, gyrators, and

*This part is based on a discussion with R. Rosenberg in November 1974, who also coined the term *portality table* as an expression of the *Paynterian spirit.*

junctions, generate or absorb energy as sources, dissipate as R-elements or R-fields, or, finally, store energy as I- and C-elements or -fields. The different elements are entered at the appropriate places in the table (which is then a kind of matrix).

The following observations apply to Table 8.4:

(1) Energy conserving means that the power fluxes add up to zero at all instants of time.

(2) Sources generate (or remove if driven in reverse) energy from the domain of interest. This is a useful simplification, but physically they transform it from another energy domain or take it from large storage elements in accordance with universal energy conservation. Sources are reversible (at least in principle).

(3) Dissipating elements take energy from the domain of interest and change it into thermal power (heat) consisting of entropy flow and temperature. This process is essentially irreversible.

(4) In R-fields we have energy dissipation over all bonds, but quite possibly power can come out from one bond if more power enters another bond.

(5) Including thermal power, the dissipating elements become power-conserving RS-fields. In a way, RS-fields shown within broken lines on Table 8.4 are needed to represent universal energy conservation with thermal effects (first law of thermodynamics). The source part is irreversible, since entropy is never destroyed.*

(6) Storage elements are reversible and return the stored energy when they are brought back to their original state. They thus conserve energy in the long run, that is in the average over several cycles. In the case of storage fields, energy can leave through a different bond.

Table 8.4 is indeed an interesting illustration of the basic physical principles operative in bond-graph elements. Inversely it also shows how such

*The exception with negative internal resistance is a simulation artifice as treated in § 6.3.

principles are exemplified by bond-graph elements. As a result, bond-graph elements are a useful tool for teaching physics.

One might be tempted to speculate how the void spaces of Table 8.4 could be filled, but there do not seem to be worth while entries. Simple bonds could also be considered as power-conserving one-ports. Junctions with only two bonds arise with systematic setup of bond graphs. They have been left out from the table since they are equivalent to simple bonds (§ 2.4).

Returning now to the different analogies, some elements change (as shown above) when going from the direct to the dual analogy. However, such exchange or permutation only takes place within the places of Table (matrix) 8.4. In other words the energy handling properties and the number of ports remain invariant under a change of analogy.

From the point of view of energy handling and of portality, there is consequently no preference between both analogies. However, at the time of writing (early 1975) it appears that the question of the interpretation of force as effort or flow has deep ramifications in the physical structure of our universe, connected presumably with the question of the absolute and relative values of physical variables, especially entropy and temperature. Further study and reflection on these lines is certain to bring fruitful results.

APPENDIX

Short Table of the Most Important Symbols

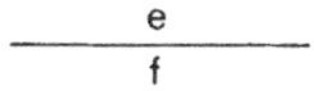

Simple bond, effort on top, flow below (or right and left).

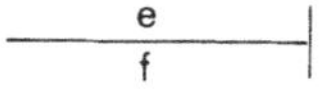

Cross-stroke for causality: effort acts on the side of the stroke; flow acts backwards.

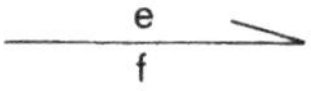

Half-arrow for positive direction of energy flux (power).

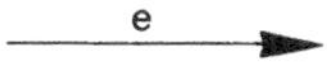

Activated bond; corresponds to a connection of a block diagram.

One-ports

 R

Resistance element (R-element) with a relation between effort and flow.

 I

Inertia element (I-element) with a relation between momentum and flux.

 C

Capacity element (C-element) with a relation between effort and displacement.

 S_e

Effort source with effort independent of flow.

 S_f

Flow source with flow independent of effort.

Observation: a control or modulation by a signal is represented by an activated bond.

Two-ports

—— TF —— Transformer (power conserving).

—— GY —— Gyrator (overcrossed transfórmer, power conserving).

Observation: If the transformer or gyrator ratio is modulated by a signal, one writes MTF and MGY.

Junctions

—— p —— Parallel junction (p-junction, power conserving), efforts equal, flows additive.

—— s —— Series junction (s-junction, power conserving), efforts additive, flows equal.

Multiports

—— R —— R-field with relations between efforts and flows.

—— I —— I-field with relations between flows and momenta.

—— C —— C-field with relations between efforts and displacements.

Observation: additional symbols are the IC-field (§ 5.3.3), the RS-field (§ 6.3), and the SS-double source (§ 6.3).

Glossary of Frequent Terms in English, French, and German

activated bond	liaison activée	aktivierte Bindung
bond graph	graphe de liaison	Bonddiagramm
one-port, two-port	monoport, biport	Einport, Zweiport
equivalent circuit	circuit équivalent	Ersatzschaltbild
gyrator	gyrateur	Gyrator
capacitance element	élément de capacité	Kapazitätselement
causality	causalité	Kausalität
multiport	multiport	Mehrport
block diagram	diagramme fonctionnel (schéma blocs)	Signalflußplan (Blockdiagramm)
effort source, flow source	source d'effort, source de flux	Spannungsquelle, Stromquelle
inheritance element	élément d'inertie	Trägheitselement
transformer	transformateur	Transformator
connection	connexion	Verbindung
junction	jonction	Verknüpfung
resistance element	élément de résistance	Widerstandselement
resistance field	champs de résistance	Widerstandsfeld
C-field	champs-C	C-Feld

References

1. Karnopp, D. and Rosenberg, R., *Analysis and Design of Multiport Systems,* MIT Press, Cambridge, USA, 1968.
2. Karnopp, D. and Rosenberg, R., *System Dynamics; A Unified Approach,* 2 vols. text, 1 vol. problems, and a computer tape in FORTRAN 4. Obtainable from Div. of Engineering Research, Michigan State University, East Lansing, Mich. 48 823, USA (to be published in 1975 as book by Wiley-Interscience, New York).
3. Different authors in special issue on bond graphs, *Trans. ASME,* **94**, Series G, No. 3 (Sept. 1972).
4. Thoma, J. and Karnopp, D., Simulation of continuous systems by diagrams, in Schöne, A. (ed.), *Simulation technischer Systeme,* Carl Hanser Verlag, Germany, 1974.
5. Thoma, J., *Modern Oil Hydraulic Engineering,* Trade and Technical Press, Morden, Surrey, 1972.
6. Thoma, J., *Einführung in die Oelhydraulik und die hydraulische Systemtechnik,* Verlag Girardet, Germany, 1973.
3. Messerle, H. K., *Energy Conversion Statics,* Academic Press, 1969.
8. Thoma, J., Bond graphs for thermal energy transport and Entropy flow, *Franklin Ist.* **292,** 109–20 (1971).
9. Job, G., *Neudarstellung der Wärmelehre,* Akademische Verlagsgesellschaft, Germany, 1972.
10. Crandal, C. and Karnopp, D., *Dynamics of Mechanical and Electromechanical Systems,* MacGraw-Hill, USA, 1968.
11. Karnopp, D., Power conserving transformations using bond graphs, *J. Franklin Inst.* **288,** 175–201 (1969).
12. Kinsler, L. and Frey, A., *Fundamentals of Acoustics,* Wiley, 1962.
13. Küpfmüller, K., *Einführung in die theoretische Elektrotechnik,* Springer Verlag, Germany, 1959.
14. Melsa, J. and Jones, S., *Computer Programs for linear Control Theory,* MacGraw-Hill, USA, 1973.
15. Cardwell, D., *From Watt to Clausius,* Heinemann, London, 1971.
16. Takahashi, Y., Rabins, M., and Auslander, D., *Control,* Addison-Wesley, USA, 1970.
17. Brouer, B., Die Bondgraph Darstellung fuer dynamische Systeme, *Chemie–Ingr–Tech.* **44,** 1022–27 (1972).

18. Karnopp, D., on the penetrating but olympian tensorial methods of Gabriel Kron, *Matrix and Tensor Quart.* **20,** (4) 120–4 (1970).
19. Benson, R., *Advanced Engineering Thermodynamics,* Pergamon Press, 1967.
20. Karnopp, D., Lumped parameter models of reactive mufflers with normal modes and bond graphs, unpublished report of the University of California at Davis, 1974.
21. Fleischmann, R., *Einführung in die Physik,* Physik Verlag, Germany, 1974. Original publication in *Zeitschrift für Physik,* **129,** 377–400, (1951) and **138,** 301–8 (1954).
22. Bell. A., and Martens, H., A comparison of linear graphs and bond graphs, *Proc. Joint Autom. Control Conf. June, 1974.*

Index

Printed in France by Amazon
Brétigny-sur-Orge, FR

51786486R00111